中国财政科学研究院智库丛书

中国气候公共支出分析与评估

——基于河北省的研究

（中英双语）

刘尚希　石英华　等著

中国财经出版传媒集团
中国财政经济出版社

图书在版编目（CIP）数据

中国气候公共支出分析与评估：基于河北省的研究/刘尚希等著．—北京：中国财政经济出版社，2017.12
（中国财政科学研究院智库丛书）
ISBN 978-7-5095-7800-1

Ⅰ.①中… Ⅱ.①刘… Ⅲ.①气候变化-财政支出-经济评价-中国
Ⅳ.①P467 ②F812.45

中国版本图书馆 CIP 数据核字（2017）第 257419 号

责任编辑：胡　博　　　　　　责任校对：刘　靖

中国财政经济出版社 出版

URL：http://www.cfeph.cn
E-mail：cfeph@cfeph.cn

（版权所有　翻印必究）

社址：北京市海淀区阜成路甲 28 号　邮政编码：100142
营销中心电话：88190406　北京财经书店电话：64033436　84041336
北京中兴印刷有限公司印刷　各地新华书店经销
787×1092 毫米　16 开　16.75 印张　300 000 字
2017 年 12 月第 1 版　2017 年 12 月北京第 1 次印刷
定价：68.00 元
ISBN 978-7-5095-7800-1
（图书出现印装问题，本社负责调换）
本社质量投诉电话：010-88190744
打击盗版举报热线：010-88190414、QQ：447268889

中国财政科学研究院智库丛书

编 委 会

编委会主任 刘尚希

编委会委员 罗文光　　白景明　　傅志华

总　　序

　　党的十八届三中全会在明确"完善和发展中国特色社会主义制度，推进国家治理体系和治理能力现代化"这一全面深化改革总目标的同时，提出了"财政是国家治理的基础和重要的支柱"的重要判断，充分彰显出财政在国家治理现代化之中的地位与作用。

　　强调发挥财政在国家治理中的基础和重要支柱作用，是与我国经济社会发展阶段相联系的。在改革开放初期，政府的作用是促进改革和开放，财政改革主要是推动政府职能转换、改进政府与市场关系，让市场在资源配置中发挥更大的作用。随着我国经济社会转型进入新的阶段、国家实力逐渐增强以及大国财政使命的提出，财政在改革和发展中的作用日趋多样化、全方位，涉及经济、政治、社会、文化、生态文明建设各个领域。

　　在市场经济不断发展的基础上，社会结构及其整个上层建筑都发生了极大变化，社会成员利益关系变得复杂起来。在经济进入新常态的背景下，这种复杂的利益关系对于财政在国家治理中作用的发挥是一个新的考验。改革开放初期，财政政策着眼于关注国内，对于国际环境关注不多，现在财政政策的一举一动都对世界经济产生重要影响；改革开放初期，财政主要解决温饱问题，经济建设成为财政工作的突出任务，现在财政既要解决发展问题，又要解决改革问题，经济、社会、政治、文化和生态文明要协同发展；改革开放初期，中央和地方财政实力虽然都较弱，但地方政府债务也少，现在国家财政实力快速扩张过程中也面临着地方政府债务特别是或有债务快速扩张的问题，财政自身可持续性发展面临挑战。

财政作为国家治理的基础正在发生多维变化。改革开放初期，财政主要从经济维度发挥国家治理基础性作用，主要是处理好政府与市场的关系；在经济社会转型、利益关系多元化背景下，财政要从多维度支撑国家治理：既有国家与市场的维度，也有国家与社会（个人）的维度，以及公共部门内部（包括中央与地方、政府部门之间）的维度。

随着财政发挥作用的多维变化，财政理念也随之发生变化。改革开放初期，政府在市场失灵的领域提供公共服务；随着时代的进步，政府承担的各种责任（城镇化、养老、医疗、教育、环境保护等）在不断增加，在政府能力有限的情况下，政府与社会资本合作呼之欲出。政府和社会资本合作打破了传统主流经济学、财政学的基本看法：政府与市场是水火不相容的，二者是对立的；公共服务领域是市场失灵的领域，只能由政府来干。过去注重政府与市场之间的分工，现阶段则注重在分工基础上的合作。政府与市场关系需要进行再改革，一些新的问题又随之产生：在多元主体提供公共服务的同时如何保障社会公共利益，如何理顺政府与社会的关系，如何理顺政府内部如中央和地方之间、政府各部门之间的关系等。财政全方位、深层次嵌入国家治理体系和治理能力现代化之中，带来了许多需要用全新理论诠释的问题，也考验着各方面的智慧。

面对新阶段、新形势和新任务，财政如何有效支撑和推动国家治理现代化更需要新思路、新思想，财政智库或财政思想库也应运而生。可以说，财政智库是财政有效支撑和推动国家治理现代化的思想源泉，也是点亮财政作用于国家治理的"智慧之灯"。发达国家在财政现代化和国家治理体系与治理能力现代化过程中，财政智库的作用功不可没。要发挥好财政作为国家治理基础与重要支柱的职能作用，财政智库的基础性作用更是不可替代。

第一，财政智库是推进国家治理决策的科学化、民主化和法制化的重要支撑。当前，全面建成小康社会进入决定性阶段，破解财政改革发展稳定难题和应对全球性问题的复杂性艰巨性前所未有，迫切需要健全中国特色的财政决策支撑体系，大力加强财政智库建设，以财政科学咨询支撑财政治理的科学决策、民主决策和依法决策，以财政科学决策引领科学发展。

第二，财政智库是国家治理体系和治理能力现代化的重要内容。纵观当今世界各国现代化发展历程，智库在国家治理中发挥着越来越重要的作用，日益成为国家治理体系中不可或缺的组成部分，是国家治理能力的重要体现。全面深化改革，推进国家治理体系和治理能力现代化，推动协商民主广泛多层制度化发展，建立更加成熟更加定型的制度体系，必须切实加强中国特色新型财政智库建设，充分发挥智库在治国理政中的重要作用。

第三，中国特色新型财政智库是国家软实力的重要组成部分。一个大国的发展进程，既是经济等硬实力提高的进程，也是思想文化等软实力提高的进程。智库是国家软实力的重要载体，越来越成为国际竞争力的重要因素，在对外交往中发挥着不可替代的作用。树立社会主义中国的良好形象，推动中华文化和当代中国价值观念走向世界，在国际舞台上发出中国声音，迫切需要发挥中国特色财政新型智库在公共外交中的重要作用，不断增强我国在国际财经和公共事务的国际影响力和国际话语权。

正是考虑到智力资源是一个国家、一个民族最宝贵的资源，考虑到我国智库发展面临的各种瓶颈，2015年1月，中共中央办公厅、国务院办公厅印发了《关于加强中国特色新型智库建设的意见》，提出加强智库建设整体规划和科学布局，统筹整合现有智库优质资源，重点建设50~100个国家急需、特色鲜明、制度创新、引领发展的专业化高端智库。

中国财政科学研究院的前身财政部财政科学研究所（财科所），于1956年根据毛泽东主席的指示而成立，2016年2月正式更名。60年前财科所成立之初，就定位为政府部门的政策咨询机构，以探索我国财政经济问题和培养财政、会计专门人才为己任，为党中央和国务院中心工作服务，为财政经济发展的现实服务。为此，一代又一代财政科研人员为我国财政科研事业做出重要贡献。60年后的今天，中国财政科学研究院正致力于转型、创新，努力创建一流新型智库。

根据智库建设与发展的规划，本院推出"中国财政科学研究院智库丛书"。该丛书内容既包括本院各年度重要《研究报告》的文集，也包括本院承担完成的一些重大科研项目成果,以及本院研究人员研究、撰写的各类专著。目的在于集中展示财科院的科研成就,扩大科研成果的宣传和社会效果,全面提升财科院的智库影响力。

不忘初心,砥砺前行。我们将明确智库建设的宗旨,在传承既有科研优势和办院特色的基础上,探寻新型高端智库建设的途径,潜心探索财政与国家治理的新理论、新观点、新思路、新对策,与各界同仁一道,共同致力于现代财政制度建设,开创国家治理现代化之美好未来。

"中国财政科学研究院智库丛书"编委会

2016年7月

中国气候公共支出分析与评估
——基于河北省的研究

课题负责人：
刘尚希（中国财政科学研究院院长，研究员）

课题组成员：
石英华（中国财政科学研究院宏观经济研究中心 主任，研究员）
潘力铭（中国财政科学研究院 博士）
罗弘毅（中国财政科学研究院 博士）

咨询专家：
高志立（河北省委常委 河北省财政厅厅长）
姚绍学（河北省财政厅副厅长）
李杰刚（河北省财政厅副厅长）
刘启生（河北省财政科学与政策研究所所长）
张　硕（河北省财政科学与政策研究所 副研究员）

评阅人：
Thomas Beloe, Governance, Climate Change Finance and Development Effectiveness Advisor, Bangkok Regional Hub (BRH), UNDP Asia Pacific Regional Centre

Sujala Pant, Senior Advisor, Bangkok Regional Hub (BRH), UNDP Asia Pacific Regional Centre

Yuan Zheng, Economist, UNDP China

序

世界正在步入以可持续发展目标（SDGs）为引领的新阶段。可持续发展目标旨在实现多维度的可持续发展，其相较千年发展目标（MDGs）而言具有更高的追求。环境可持续性在确保经济持续、包容和平衡的增长方面有重要作用，因此正受到来自各方越来越多的关注。为了达成应对气候变化、保护生物多样性和实现自然资源可持续利用等目标，社会各界需要共同努力。

中国高度重视应对气候变化。中国政府已经制定了一系列目标来指导这方面的行动与投资。例如，中国在2016年提交了应对气候变化文件《国家自主贡献（Intended Nationally Determined Contribution，简称INDC）》，确立了国内层面的减排承诺。到2030年，中国计划将其单位国内生产总值二氧化碳排放量比2005年降低60%—65%[①]；二氧化碳排放2030年左右达到峰值并争取尽早达峰。同时，中国在气候变化领域的南南合作与日俱增[②]。紧紧跟随全球经济治理不断变化的国际形势，中国承诺设立200亿人民币（约30亿美元）的气候变化南南合作基金，在国际范围内取得了重要的领导地位。

国际社会高度期待中国在推动绿色增长方面发挥开创性的作用。中国领导层已将许多有关绿色发展的重要发展战略写入了《中华人民共和国国民经济和社会发展第十三个五年规划纲要》（简称"十三五"规划）之中。

1. 生态文明建设

"十三五"规划包含创新、协调、绿色、开放、共享五大发展理念。因

[①] 中国在《国家自主贡献》中强调的目标可参见：http://www4.unfccc.int/submissions/INDC/Published%20Documents/China/1/China's%20INDC%20-%20on%202030%20June%202015.pdf.

[②] 参见：https://eneken.ieej.or.jp/data/6813.pdf.

此，绿色增长成为2020年总体发展指导原则之一。根据"十三五"规划中的相关理念，以加快绿色增长为目标，中国正在全力建设"生态文明"。"生态文明"这个概念首次于中国共产党第十七次全国代表大会报告中提出，并于十八大时列为"五位一体"战略发展格局的重要要素之一，取得了同经济、政治、文化和社会发展同等的重视程度。

"生态文明"这个概念源于中国哲学"天人合一"的理念。生态文明的定义已被广泛讨论，可以从以下两个角度理解①。其一，从时间维度来看，生态文明被认为是工业文明基础之上的人类新的文明形态。这个观点关注人与自然之间的平衡，认为社会发展是依赖自然环境的，应当在尊重、保护自然的前提下寻求最大限度的发展，形成人与自然的和谐、一体，而不是对立、冲突。其二，从资源禀赋的角度看，生态文明被认为是由几个要素组成的总体架构。这意味着，建设生态文明需要多维度调整生产消费模式、制度建设和法制规范等。

总而言之，生态文明是"以尊重自然承载力的方式来实现人类繁荣"的观念②。它促进自然资源的有效利用，低碳发展以及自然环境的安全和健康。这在很大程度上符合可持续发展的各项原则及在其基础上提出的可持续发展目标。

"十三五"规划包含以绿色增长为中心的一系列目标。这些目标可被主要分为两组③。一组目标涉及相对绿色增长，重点关注资源/能源的利用效率，另一组则是绝对绿色增长，关注总资源/能源消耗（见表1）。

表1   "十三五"规划主要绿色增长目标大纲

分组	目标
相对绿色增长	- 到2020年，每单位GDP的能源消费减少15%（对比2015的数据） - 到2020年，每单位GDP的二氧化碳排放量减少18%（对比2015的数据）④
绝对绿色增长	- 到2020年，用水总量控制在6700亿立方米（对比2015年数据⑤） - 2020年的总基础能源消费小于50亿吨标准煤

① 郭濂等编著：《生态文明建设与深化绿色金融实践》，中国金融出版社。
② 2016《南南合作有关生态文明建设》，中国环境与发展国际合作委员会。
③ 中国"十三五"期间绿色增长路线图，2015，中国环境保护部，环境与经济全球绿色增长研究所和政策研究中心。
④ 这个目标同中国应对气候变化目标相一致（2014—2020）。
⑤ 详见"十三五"规划：http://www.miit.gov.cn/n1146290/n1146392/c4676365/content.html。

此外，政府还提出了多元化的政策工具和措施，旨在为绿色增长创造有利环境。例如，到2020年，中国将建立由八个制度组成的体制框架用以促进生态进步，包括自然资源的产权制度、发展及保护国土等①。为加强环境治理，中国自2015年以来修订实施了《中华人民共和国环境保护法》和《中华人民共和国大气污染防治法》。在"十三五"期间，以这些法律为中心建立起一套责任体系，防止地方政府不当干预，鼓励公众参与承担减缓气候变化的责任。

又如，"十三五"期间开展了空气、水和土地污染控制等多项规划②。在污染控制方面，"十三五"文件明确指出，政府将开展1000万亩受污染耕地治理修复和4000万亩受污染耕地风险管控③。关于废物处理措施，将建设5个低放射性废物处置场和一个高放射性废物处置地下实验室。为了进一步保护自然环境和完善绿色生态建设，政府将建立27万平方公里的土壤侵蚀治理区，以及不低于8亿亩的国家湿地。总体来看，共10项防治土地污染的措施即将实施，同时已有20余项有关空气和水污染的措施生效。

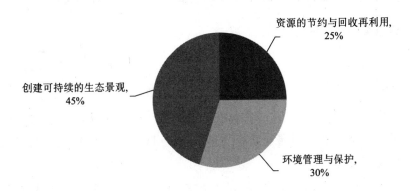

图1　"十三五"规划有关绿色发展的措施

资料来源：《中华人民共和国国民经济和社会发展第十三个五年规划纲要》。

2. 气候公共财政统计分析研究

在越来越多以推进中国绿色发展为前提的政策手段和措施陆续出台的背

---

① 详情请见：http://news.xinhuanet.com/english/china/2015-09/21/c_134646023.htm。
② "十三五"规划详情请见：http://en.ndrc.gov.cn/newsrelease/201612/P020161207645765233498.pdf。
③ 来自新华社报道：http://en.xfafinance.com/html/Policy/2015/155187.shtml。

景下，是否存在相关的机构设置，以确保政策得到妥善执行？以绿色增长为目标的项目是否被优先考虑，从而保证有足够的资金支持？是否已经为战略性的绿色发展计划进行了预算规划？绿色融资是否满足绿色投资的需求？以上这些问题对以事实和证据为依据的政策制定都至关重要。

《中国气候公共支出分析与评估——基于河北省的研究》提供了一个可以系统审查一国的制度、政策组合，公共支出及其与气候相关活动密切程度的机会。气候公共支出制度评估（CPEIR）研究由联合国开发计划署亚太区域中心发起，自2011年以来在20多个国家相继开展。该研究是一个创新的工具，以期为全面气候融资框架在国别和地方层级的建立打造坚实的基础。其研究结果和建议力求将气候变化因素纳入预算制定的项目考量之中，为支持公共财政"漂绿"而努力。

在中国，此研究目的在于更好地理解促进生态文明建设相关的各项融资、制度及政策，从而扩大有效决策范围，更有效地促进变革。这项实践希望可以为促进绿色金融在内的各项绿色发展提供有力见解和支撑。

3. 绿色金融在中国

中国近年来一直大力推动投资向绿色发展转型。"十一五"期间（2006—2010年），中国在新能源行业投资约2560亿美元，在提高能源效率方面投资达到1270亿美元[①]。2012年，中央财政投入979亿元（约合142亿美元）用于节能减排和可再生能源专项资金，以及诸如农业、水利、海洋管理、卫生和气象等行业[②]。

然而，考虑到可持续能源发展、能源和资源使用效率、环境整治和保护、污染治理以及绿化建设等项目的投资需求，中国在环境治理相关领域的融资需求显著。据中国人民银行2015年初的估计，在未来5年时间里，中国政府每年需投资约3%的GDP解决气候、水和土地的问题[③]。

2016年3月，"十三五"规划明确提出"建立绿色金融体系，促进绿色

---

① 参见：http://newenergynews.blogspot.tw/2013/03/todays-study-chinas-new-energy.html。
② 参见：http://www.sckxzx.com/index.php?_m=mod_article&_a=article_content&article_id=98。
③ 《构建中国绿色金融体系》，绿色金融工作小组，2015年。

信贷发展和建立绿色发展基金"①。目前来说，绿色信贷仍然是绿色金融的主要来源。截至 2014 年底，根据中国银行业监督管理委员会的统计，中国 21 个主要银行总共发放绿色贷款约 6 万亿人民币（约合 1 万亿美元）②。2016 年，中国已成为世界最大的绿色债权市场，市场总额达到 340 亿美元③。伴随着对上市公司环境信息的披露，绿色证券指数的发布和绿色投资基金的成立，中国绿色股票市场取得了长足的发展。此外，中国的保险市场已经引入强制性环境责任保险。但是这项试点的结果并不令人满意，因为只有少量的保险公司和企业参与其中。

引导绿色金融的另一个重要手段是中国的碳排放交易体系。碳排放权交易于 2013 年开始在以下 7 个省市运行：北京、天津、上海、重庆、湖北、广东和深圳④。自 2013 年到 2015 年，碳市场交易值达到 11 亿元人民币（约合 1.6 亿美元），减排 68.6 万吨二氧化碳⑤。作为改革的进一步标志，2017 年全国层面将启动碳排放交易体系。开发创新性的碳排放金融交易产品将会在与金融和中介机构的积极合作中进一步开发。

中国人民银行和其他七个部委于 2016 年 8 月 31 日正式发布了《关于构建绿色金融体系的指导意见》⑥。其目标是为中国绿色金融全面铺路，按照传统的绿色信贷、债券、股权和保险产品建立税收补贴和商业银行绿色补贴⑦。拟议的激励措施包括实行差别化的存款准备金率，放松对补充按揭贷款的管制，并允许一定权重的高风险贷款。另外，中国将建立首个异地碳排放衍生品交易业务，并且在金融机构层面推动环境压力测试，旨在不断加强绿色金融领域的国际合作⑧。

---

① "十三五"规划参见：http：//www.china.com.cn/lianghui/news/2016－03/17/content_38053101_13.htm。
② 绿色金融在中国银行业蓬勃发展：http：//www.chinadaily.com.cn/bizchina/greenchina/2015－08/26/content_21709767.htm。
③ Green finance progress report, 2017, UN Environment。
④ 来自 CCICED：http：//www.cciced.net/cciceden/PublicationsDownload/201702/P020170210473297581978.pdf。
⑤ 来自 The Climate Group：https：//www.theclimategroup.org/news/spotlight－china－new－emissions－trading－system－set－revamp－global－market。
⑥ 《关于构建绿色金融体系的指导意见》，参见：http：//gongwen.cnrencai.com/yijian/92291.html。
⑦ 参见：http：//news.163.com/16/0229/02/BGV7K4LA00014AED.html。
⑧ 参见：http：//www.nbd.com.cn/articles/2017－01－12/1069585.html。

4. "一带一路"与绿色发展

"一带一路"是由中国牵头的区域发展倡议。以互联互通为主题,"一带一路"旨在通过在经济、社会、政策、金融和电子等方面的有效连接,实现空间资源最优分配与整合,推动区域合作和对话,实现互惠互利。"一带一路"倡议自提出以来,得到了国际社会的广泛关注。其原因之一是该倡议有巨大潜力,可为可持续发展做出长期贡献,创造有用的价值。已有研究表明,"一带一路"倡议的原则和主要发展领域与可持续发展目标在多方面高度一致①。若能实现两者的有效对接,"一带一路"将在可持续发展领域帮助各国向自己的既定目标不断迈进。

在绿色发展领域,"一带一路"蕴藏着很多机遇。绿色发展可通过多种途径实现,例如绿色建筑,绿色基础设施建设等。机会不仅限于此。绿色投资、绿色贸易及绿色技术转移都可实现南南合作在绿色发展中的突破。绿色金融体系在相关国家的建立更是潜在的助力。因此,如何进一步开发"一带一路"的潜力,使其成为区域绿色发展的加速器,是亟待研究的课题。

5. 关于本报告

联合国开发计划署与中国财政科学研究院(原财政部财政科学研究所)于2014年至2015年间首次合作,做了课题《中国气候公共财政统计分析研究》。第一阶段的报告量化了中央政府一级气候相关公共支出。这项研究同时回顾了与气候变化和财政分配相关的关键体制框架。此研究还同时涉及其他发展中国家气候治理经验的梳理,并总结了其对中国有用的经验教训。

报告结果显示,尽管中国取得了一定程度的成功,但在环境治理方面的公共支出呈下降趋势。国家政府在节能减排和环境保护方面的支出占比从2010年的2.7%下降到2013年的1.3%。此外,报告还发现在2014年中央政府预算的7%用于直接或部分针对解决气候变化的问题(见图2)。

---

① Pingfan Hong, 2016, " Jointly building the 'Belt and Road' towards the Sustainable Development Goals", United Nations Department of Economic and Social Affairs.

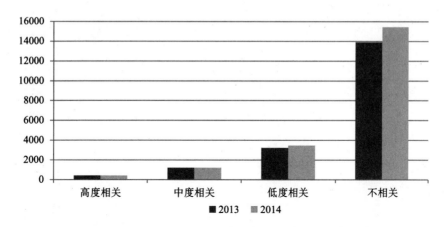

图 2　中央公共支出气候相关性（单位：亿元人民币）

资料来源：《中国气候公共财政统计分析研究》。

鉴于第一阶段研究成果，联合国开发计划署和中国财政科学研究院共同启动了第二阶段的研究，聚焦于省级层面气候公共支出分析与评估。中央的政策最终都需要本地化才能生效，地方财政支出会对政策决策起至关重要的作用。因此，省级层面的研究可以帮助更好地了解省级气候行动治理组织制度结构和气候相关支出规模，为预算规划提供有用的信息支持。在此基础上，本报告将初步对气候相关财政支出进行成本效益分析。这项工作意在识别更好的资金管理办法及其影响最大化的手段切入点。

具体来说，报告旨在：（1）在省级层面建立气候相关财政支出的基准线；（2）评估气候相关活动的公共支出与省级层面气候变化或绿色增长的政策优先事项和战略的吻合程度；（3）辨识实现环境可持续发展的瓶颈和机遇。

联合国开发计划署

2017 年 7 月 31 日

前　　言

　　气候问题已成为日益严峻的国际问题，应对气候变化是国际社会的共同任务。中国实施了生态文明建设战略，积极参与全球应对气候变化、推动可持续发展的行动。2016年，中国财科院承担了联合国开发计划署（UNDP）资助的"气候公共支出与制度评估"（Climate Public Expenditure and Institutional Review，CPEIR）第二期研究，这是UNDP继2014—2015年中国财科院专家圆满完成CPEIR第一期研究后的持续资助研究项目。CPEIR第一期研究构建了中国气候公共财政统计方法，并对中央本级气候公共支出进行统计分析。研究成果得到联合国相关机构、财政部、发改委相关司局的高度认可。第二期研究向纵深推进，重点关注中国地方层级气候公共支出与制度评估，通过案例研究进行中国省级气候公共支出与制度评估，探讨气候支出的成本—效益分析方法。

　　项目承担单位中国财政科学研究院高度重视气候公共支出研究，成立了以刘尚希院长牵头的课题组。与此同时，本项目负责人刘尚希院长也当选为第三届国家气候变化专家委员会委员，可以更好地推动本项目的研究。另外，邀请河北省财政厅高志立厅长（现为河北省委常委）、姚绍学副厅长、李杰刚副厅长、刘启生所长、张硕副研究员等担任项目咨询专家。项目于2016年7月初步确定研究任务，即以河北省为案例开展中国省级气候公共支出与制度评估研究，并选取具有典型性的"去产能"案例进行气候支出成本—效益分析的研究。项目研究期间，项目组做了大量的文献、数据搜集工作，召开课题启动会，两次赴河北省作深度调研，定期与UNDP专家讨论交流，与国际专家探讨交流。

　　2016年10月10日，项目组在河北省石家庄市举办了"省级气候变化公共支出与制度评估"项目启动会,来自河北省财政厅、水利厅、林业厅、卫计委、环保厅、发改委、工信厅、农业厅、气象局等部门的专家、中国财政科学研究院课题组专家、联合国开发计划署驻华代表处项目官员等参加了会议。会上,项目负责人中国财政科学研究院院长刘尚希研究员、河北省财政厅姚绍学副厅长、UNDP驻华代表处郑元博士分别致辞。石英华研究员代表课题组介绍了课题研究背景、研究思路和拟采用的方法。与会专家围绕省级气候公共支出评估研究发表了各自见解和看法。

刘尚希院长指出,本项研究是评价财政支出对减缓、适应气候变化的作用与成效。气候公共支出是财政支出结构优化的重要部分,支出是否在向有利于减缓、适应气候变化的方向演化,是衡量其是否优化的关键。气候变化对人类的生存、发展有重大影响,研究极具价值,中国有责任、有义务探索出一条应对气候变化的新路来,这也是中国作为大国所应该做的。做全国的研究,工作量非常大。河北省是京津冀一体化的重要成员,在体制改革方面敢于突破,在应对气候变化方面做了很多工作,选取河北作为案例分析,能够以点带面。联合国开发计划署特别支持本项研究,也认可选取河北作为案例研究。本项研究要从理念、政策、体制机制角度,总结经验,既不能满足现有的做法,也要提取优秀经验,作为发展中国家的样板。

在随后进行的第一次调研座谈会上,项目组与河北省水利厅、林业厅、卫计委、环保厅、发改委、工信厅、农业厅、气象局等省直单位的领导和专家进行了热烈的讨论和交流,并与省财政厅预算局、经建处、资环处、农业处、综合处、税政处、社保处、PPP办、采购办、科研所等气候相关各职能部门专家就气候公共支出及预算制度进行了深入探讨。

鉴于去产能是近年来河北省应对气候变化的重要举措之一,项目组选取河北省"去产能"项目进行研究。2017年4月24日到26日,项目组由刘尚希院长带队赴河北省进行气候公共支出成本收益分析的案例调研,河北省财政厅李杰刚副厅长全程参与指导调研。调研期间,项目组与河北省发改委、财政厅、工业信息厅、环保厅、人力社会保障厅等省直单位,省冶金协会、省煤炭协会等行业协会,石家庄市及下属平山县工业信息局,唐山市发改委、财政局、工信局、人社局、环保局等市直单位,以及丰南区发改委、财政局、人社局、政府办、工信局等区直单位的领导和专家进行了深入的座谈交流,实地考察了河钢集团最大的子公司唐钢、贝钢等企业以及平山县一家已关停的水泥企业，了解河北省去产能的进展、产生的效益以及面临的挑战和问题。

2016年9月5日到9日，研究项目组成员中国财科院宏观经济研究中心主任石英华研究员、河北省财政科研所张硕副研究员、UNDP驻华代表处郑元博士赴巴基斯坦参加了"气候变化财政、规划和预算制度的融合"（Integrating Climate Change Finance in Planning and Budgeting Systems）双边交流。在巴期间，项目组与巴基斯坦顶级智库"环境与发展领导力"（Leadership for Environment and Development，LEAD）就气候公共支出与制度评估的目标、要素、方法、数据获得、测度指标、时间敏感性等方面进行了交流；拜会了巴基斯坦财政部部长助理Syed Ghazanfar Abbas Jilani先生，了解巴政府部门对CPEIR成果的应用情况，还拜访了正在巴访问的UNDP曼谷大区CPEIR项目主管Sujala Pant女士以及UNDP驻巴代表处、巴气候变化部、总会计署等，参加了LEAD在拉合尔市举办的旁遮普省CPEIR启动会。此次双边交流有助于加深对CPEIR方法的理解和共识，学习成本效益和物有所值分析方法在气候支出方面的应用，以推动我国气候公共支出评估工作的开展。

项目研究产生了良好的效果。项目组专家多次作为专家参加相关气候变化公共支出政策研讨咨询。2016年9月30日，本项目负责人刘尚希院长参加第三届国家气候变化专家委员会成立大会，以及之后的不定期研讨会。项目组完成了《中国省级层面气候公共支出与制度评估——以河北省为例》、《气候公共支出项目的成本效益分析——以河北省"去产能"为例》两份研究报告，调研后形成的《去产能不应是目标》、《关于去产能与降成本、去杠杆、增利润关系的进一步分析》等调研成果以《简报》、《专报》等形式报送相关政府决策部门，并受到财政部肖捷部长的高度重视和肯定。

2017年9月8日，中国财政科学研究院与联合国开发计划署在京举行气候公共支出分析与评估研究成果发布暨研讨会。联合国开发计划署驻华代表处国别主任文霭洁（Agi Veres）、联合国开发计划署亚太局高级区域顾问Asad Maken、环保部政策法规司司长别涛、国家行政学院中国生态文明研究中心主任张孝德教授、河北省财政厅副厅长李杰刚、世界资源研究所高级研究员朱寿庆、发改委应对气候变化司综合处处长黄问航、工信部节能司综合处副处长郭丰源、环保部经济政策研究中心副研究员刘哲等与会专家对项目研究的理论价值、现实意义、研究方法、分析框架等给予高度评价。

项目研究期间，我们得到了河北省相关部门、行业协会、企业的大力支持、指导和帮助。衷心感谢河北省财政厅的各位领导和专家对项目研究给予的有价值的建议和十分有力的帮助。感谢河北省相关省直单位和行业协会，石家庄市及平山县工信局，唐山市市直单位，以及丰南区区直单位的积极参与。还要感谢我们实地调研中给予热情帮助的各个企业。感谢 UNDP 专家 Niels Knudsen 和 Carsten Germer 对报告的审阅及对项目的积极支持和悉心指导。感谢 UNDP 的研究助理卫栎及 Andrew Cheng 对于本报告的贡献。

气候公共支出评估研究是个崭新的领域，目前对气候公共支出的统计口径和范围的界定尚待统一，评估方法尚待完善，进行气候公共支出评估研究富有挑战性，本项研究反映了我们的初步探讨。今后，我们将继续探索，为人类应对气候变化尽绵薄之力。

中国财科院《中国气候公共支出分析与评估——基于河北省的研究》课题组
2017 年 9 月 30 日

内容提要

气候问题已成为日益严峻的国际问题和全球可持续发展的重大问题。中国站在人类文明发展的高度,提出"生态文明"的新理念,并以此为指导,积极应对气候变化、维护全球生态安全。中国已将应对气候变化纳入国民经济和社会发展规划,地方政府也制定了大量的政策法规以应对气候变化,并落实到公共支出。省级政府在应对气候变化方面扮演了重要角色,研究省级层面气候公共支出与制度评估具有重要意义,有助于有效刻画中国在生态文明建设方面的生动实践,探索出一条独特的生态文明发展之路,有利于支持应对气候变化决策,有利于中国地方与国外气候财政支出的比较及交流。

河北省地处华北平原,内环京津、外沿渤海,在应对气候变化方面,河北省具有典型性。河北是工业大省,钢铁产量占全国产量的1/4,是全国资源环境与发展矛盾较为突出的地区之一,环境治理面临的压力与挑战极大。近年来,河北省积极应对气候变化,不断增加气候相关公共支出,应对气候变化取得了一定成效。选取河北作为中国地方政府应对气候变化公共支出政策的案例作分析,能够以点带面,把握地方政府整体支出情况,也可提取优秀经验,促进地方政府有效应对气候变化。

一、河北省气候公共支出与制度评估

通过评估河北省气候公共支出与制度,本报告得出如下结论:

1. 气候变化问题日益受到重视,气候公共支出逐年增长。评估结果显示,"十二五"期间,气候公共支出逐年增长。尤其是2015年以来河北省加大节能环保等方面的投入,与气候变化高度相关支出较上年增长达38.7%。根据气候支出的相关度分析,2011—2015年,与气候高度、中度相关的财政

支出整体呈上升趋势，全省财政支出中与气候变化高度相关与中度相关支出合计占比从8.23%上升到11.29%。从气候公共支出在全省财政支出的占比来看，河北省的气候公共支出占比高于中央气候相关支出的同口径数据。

2. 气候公共支出统计评估方法得以拓展，气候财政的方法研究得以深化。在中国气候公共支出与制度评估项目第一阶段，仅按照与应对气候变化的相关度高低分类统计气候财政资金，评估中央层面应对气候变化的公共支出情况；项目第二阶段，本报告对气候公共支出分类统计方法作了拓展，在按照气候变化相关度对河北省气候公共支出分类统计的基础上，又按照减缓气候变化与适应气候变化两个维度，对河北省气候变化财政资金按照相关度作了统计分析。本报告发现，对气候变化公共支出按照减缓气候变化与适应气候变化活动的相关度二次细分的统计分析方法，可以更加精确地统计用于应对气候变化的公共支出数额，为公共支出的成本效益分析与绩效评价提供基础。

结合气候公共支出与制度评估结论，本报告提出如下政策建议：（1）应对气候变化，应强化理念渗透和政策的顶层设计，形成政府与市场、政府与社会的良性互动；（2）继续深化预算管理制度改革，进一步提高政策和资金的有效性；（3）进一步创新财政支出方式，建立市场化的约束机制；（4）加强气候公共支出成本效益评估的研究和指引。

二、河北省去产能的成本效益分析

气候变化问题与经济发展方式密切相关。促进经济发展方式向低碳转型是应对气候变化的必然选择。当前，中国正在推进供给侧结构性改革，去产能是供给侧结构性改革的重要内容。实施去产能，能够优化升级生产结构，提升资源利用效率，降低工业排放，提高经济发展质量，是应对气候变化的重要举措。鉴于去产能是近年来河北省应对气候变化的重要举措之一，本报告选取河北省"去产能"项目作案例研究，结合对河北省相关政府部门、企业的实地走访调研，较为系统地分析了河北省去产能的成本与收益，探索构建了气候公共支出成本收益分析的方法框架。

通过对去产能项目的成本效益分析，本报告得出如下结论：

1. 去产能是应对气候变化的重要措施，相关研究亟待加强。当前及今后，中国促进经济发展方式向低碳绿色转型是应对气候变化的必然选择。

淘汰落后产能是促进经济发展方式转变，提升经济发展质量，应对气候变化的重要措施。去产能的制度、政策、工具、成本效益评估等相关研究亟待加强。

2. "去产能"的高昂成本由政府、企业等相关利益主体分担。本报告根据各利益主体及其活动特点，采用显性成本、隐性成本、机会成本的概念与分类，以去产能中的利益主体作为纵轴，以成本分类作为横轴，以矩阵形式呈现"去产能"成本。多维度矩阵分析表明，去产能过程中，政府、企业、银行等各利益相关方付出了巨大的成本代价。在总体成本的分析中，除了关注中央政府、河北省政府以及省内的县市政府安排专项资金支持去产能的显性支出外，还应关注各级政府、企业、银行等相关主体承担的各种隐性支出，尽管后者很难做出精确的量化分析，但这些支出应得到关注。

3. 去产能在实现经济、社会、生态可持续方面取得明显成效。"去产能"旨在通过工业领域生产方式的变革，实现由粗放经济向循环经济的转变，减少工业废弃物对环境的污染与破坏，降低人类活动对气候的负面影响。"去产能"是生态文明建设的重要组成部分，对改善居民的生活环境、对气候改善有重要意义。去产能可产生良好的社会效益和生态效益，通过改善环境质量，促进居民健康。通过资源合理利用，去产能可以实现社会可持续发展，有利于实现居民就业的长期稳定，实现居民收入的稳步提升，保障民生发展。去产能可以产生明显的经济效益，促进经济发展质量提升，促进经济转型升级发展，促进产业结构和产业布局的优化。

4. 从短期看，去产能的成本大于收益；从长期看，去产能的收益大于成本。结合对河北省企业实地调研情况以及本报告对成本收益的分析，从短期看，去产能的成本大于收益，政府、企业、银行等相关利益主体去产能当期的支出压力较大。对河北省去产能行业负债、成本、利润数据的分析表明，去产能导致相关行业企业当前生产成本率提升和利润率下降。但是，从长远看，去产能的收益大于成本。微观层面看，去产能有助于企业降低成本和增加利润。通过提高环保、能耗、质量、安全等生产指标去产能，能够有效剔除落后产能，降低社会整体生产成本。通过调整生产结构，提高生产附加值，最终推动企业利润平稳上升。从宏观层面看，去产能有助于实现经济、社会、生态可持续。

结合去产能项目的成本效益分析，本报告提出如下建议：（1）去产能需与发展循环经济、优化产业布局统筹考虑；（2）应分类施策，动态优化去产能政策；（3）中央政府实施的奖补资金政策应予完善；（4）气候公共支出成本收益的多维度矩阵分析框架还有待进一步拓展和完善；（5）气候公共支出相关基础信息数据的采集和统计工作亟待加强。

目录 ■■■■

报告一：中国省级层面气候公共支出与制度评估——以河北省为例
....（23）

  一、进行地方层面气候公共支出与制度评估的必要性 ....（23）
    （一）省级政府在应对气候变化方面扮演了重要角色 ....（24）
    （二）公共支出是省级政府应对气候变化的重要工具 ....（24）
    （三）省级层面气候公共支出与制度评估的重要意义 ....（26）
  二、河北省气候变化与公共政策及管理现状 ....（27）
    （一）河北省的气候变化现状 ....（27）
    （二）规划与法规制定 ....（28）
    （三）河北省气候变化管理机构 ....（33）
  三、河北省气候公共支出预算制度 ....（34）
    （一）预算管理层级 ....（34）
    （二）预算编制 ....（34）
    （三）预算科目设置 ....（35）
  四、河北省气候公共支出统计与评估 ....（37）
    （一）河北省气候公共财政统计范围的确定 ....（37）
    （二）河北省气候公共支出统计 ....（37）
  五、主要结论与建议 ....（45）
    （一）主要结论 ....（45）
    （二）相关建议 ....（46）

报告二：气候公共支出的成本效益分析——以河北省"去产能"为例
....（49）

  引言 ....（49）
  一、河北省去产能进展 ....（50）
    （一）河北去产能行业基本情况 ....（50）

（二）近年河北去产能进展 …………………………………（50）
　　（三）未来河北去产能规划 …………………………………（51）
二、去产能的成本分析 …………………………………………（52）
　　（一）去产能成本分析的理论框架 …………………………（52）
　　（二）政府去产能的成本分析 ………………………………（56）
　　（三）企业去产能成本分析 …………………………………（61）
　　（四）金融系统去产能成本分析（以银行为例）……………（62）
三、去产能的效益分析 …………………………………………（63）
　　（一）生态可持续分析 ………………………………………（63）
　　（二）社会可持续分析 ………………………………………（66）
　　（三）经济可持续分析 ………………………………………（67）
四、主要结论与建议 ……………………………………………（71）
　　（一）主要结论 ………………………………………………（71）
　　（二）相关建议 ………………………………………………（72）

结论 ……………………………………………………………（74）

附件（一）河北省气候相关公共支出政策 …………………（77）

附件（二）河北省气候相关政府部门 ………………………（84）

附件（三）近年河北省财政预算管理改革的主要内容 ……（89）

附件（四）河北省一般公共预算支出与气候的相关度分类分析表 ……（93）

**附件（五）河北省一般公共预算支出根据减缓和适应气候变化相关度
　　　　　　细分类表** …………………………………………（102）

参考文献 ………………………………………………………（111）

报告一：中国省级层面气候公共支出与制度评估
——以河北省为例

一、进行地方层面气候公共支出与制度评估的必要性

气候变化及其影响已引起世界范围的广泛关注。联合国政府间气候变化专门委员会（IPCC）第五次评估报告指出，2000年至2010年间，人为温室气体排放量平均每年增长2.2%，高于此前30年1.3%的年均增长率。20世纪以来，全球海平面上升19厘米，平均每年上升1.7毫米，人类生存环境正在以惊人的速度发生改变。温度升高、海平面上升、极端气候事件频发等诸多事件给人类生存和发展带来严峻挑战。IPCC报告指出，21世纪以来，因全球变暖导致海平面升高而引发的自然灾害造成的经济损失已高达2.5万亿美元，到2050年，每年造成的损失预计将超过1万亿美元。气候变化还可通过热应激反应、加速传染病传播、恶化人类生存条件等对人类健康直接造成危害[①]。

中国将加快推进生态文明建设作为积极应对气候变化、维护全球生态安全的重大举措，并已将应对气候变化纳入国民经济和社会发展规划，地方政府也制定了大量的政策法规以妥善应对气候变化。公共支出是地方政府应对气候变化的重要工具，对省级层面的气候公共支出与制度进行评估以优化支出安排，显得尤为重要。

[①] 许慧慧、施烨闻、钱海雷、金奇昂、张莉君、张江华、郭常义：《气候变化对人类健康的影响》，2011年全国环境卫生学术年会论文，2011年。

（一）省级政府在应对气候变化方面扮演了重要角色

1. 国家应对气候变化需要通过省级政府落实

从学理上来看，地方实施有利于缓解信息不对称问题。虽然中央决策有利于克服气候变化行动的外部性，维护统一市场，避免囚徒困境的产生，但考虑到信息不对称问题，地方政府在实际操作层面较中央的信息优势更突出，具体事务由地方落实有利于降低成本。

从现有的资源配置情况来看，地方是主要的配置主体。仅以节能环保、农林水、国土海洋气象、医疗卫生这四类与气候变化高度相关的预算支出为例，2015年全国上述四类支出总额为36251.26亿元，其中中央支出1571.64亿元，仅占全国支出的4.34%，地方支出占比则达到95%以上，具有显著重要性。

2. 妥善应对气候变化是地方可持续发展的关键

一方面，气候环境问题已经严重影响了地方经济社会发展。气候环境变化使得极端气候频发、破坏力增强，内涝、洪水、干旱、台风、风雹、低温冰冻、雪灾等气候条件严重影响了人民生产生活。据国家减灾办公布数据，2015年自然灾害致中国1.86亿人次受灾，损失2704亿元。气候环境变化已成为地方可持续发展最重大的威胁，地方政府必须妥善应对气候变化，实现绿色低碳可持续发展。

另一方面，积极应对气候变化是促进地方经济社会发展转型的重大机遇和重要抓手。中国作为一个资源禀赋较差、人均收入不高、贫困人口众多的国家，面临着发展经济、消除贫困、改善民生、保护环境、应对气候变化等多重挑战。原有发展模式下，中国生产和消费需要的资源能源已经超过生态承载条件，资源环境压力不断加强、难以承受、不可持续，必须要积极应对气候变化，提高资源节约力度，化解资源环境压力，用最少的资源能源消耗实现经济增长和满足消费需求。

（二）公共支出是省级政府应对气候变化的重要工具

1. 气候的外部性要求财政支持

一般认为，气候变化是指世界范围内温度的上升和风暴活动的增加等，是气候平均状态统计学意义上的巨大改变或者持续较长一段时间的气候变动趋势。尽管引起气候变化的原因可能是自然内部的因素，也可能是外界强迫或者人为造成的，但现有研究表明，气候类型的长期改变主要是由人类排放二氧化碳等温室气体行为所引起的。《联合国气候变化框架公约》（UNFCCC）第1款中，就将"气候变化"定义为："经过相当一段时间的观察，在自然气候变化之外由人类活动直接或间接地改变全球大气组成所导致的气候改变"。

气候变化既是环境问题，也是发展问题。由于环境具有非排他性和有限的非竞争性，因而环境资源属于公共产品。气候变化作为全球性普遍关注的重要环境问题，也应属于公共产品范畴。气候变化具有明显的外部性特征。外部性的存在导致私人边际成本与社会边际成本或私人边际收益与社会边际收益不一致，从而扭曲了价格信号，使得基于市场竞争机制达成的产品或服务市场均衡并非帕累托最优。在气候变化中的正、负外部性都将影响资源配置的效率，导致市场失灵。市场失灵需要政府干预和财政支持。

2. 财政的公共性决定了支持气候的必要性

世界银行在其1997年的世界发展报告中将每一个政府的核心使命概括为五项最基本的责任，大体上反映了现代政府所行使的职能，包括：确定法律基础；保持一个未被破坏的政策环境，包括保持宏观经济的稳定；投资于基本的社会服务和社会基础设施；保护弱势群体；保护环境。气候是人类赖以生存的重要环境内容之一，应对气候变化则是政府履行环境保护职责的组成部分。由于气候变化所具有的全球性公共产品属性及其显著的外部性特点，导致普遍存在市场失灵现象，这就决定了政府必须介入，承担起应对气候变化的职能。从全球应对气候变化的实践也可以看到，政府是推动应对气候变化的领导者和组织者。而财政作为国家治理的基础和重要支柱，有必要积极且广泛地参与到应对气候变化工作中来①。

3. 财政通过多种途径引导气候应对支出

（1）财政政策是应对气候变化的重要政策工具

作为政府实行经济调控的重要政策手段之一，财政政策对于完善市场机制，促进社会经济的综合、协调发展有着重要影响。特别是在应对气候变化领域，由于气候变化和应对气候变化具有公共产品属性，其外部性造成的市场失灵必须依赖政策手段加以解决，实施财政政策加以引导的必要性更加显著。财政政策一直是各国政府促进应对气候变化的重要政策工具。同时，通过财政政策与金融、产业等政策的协调与配合，对于应对和适应气候变化，具有重要的意义。

（2）财政政策是应对气候变化的物质保障

应对气候变化需要有大量的资金投入，考虑到气候变化的不确定性和较大风

---

① 财政部财政科学研究所课题组（苏明等）：《中国气候公共财政统计分析研究》，2015年3月。

险，私人资本较少涉足这一领域，政府的财政投入是应对气候变化的主要资金来源。具体来看，税收是政府财政收入的主要来源，是政府提供公共产品、履行公共职能的重要保障。税收在促进绿色增长中除了具有激励约束功能外，还具有筹集环保资金的功能。通过开征碳税，完善环境资源税、消费税，进一步扩大环保税基，可以为气候变化提供稳定而充足的资金来源。同时，公共财政对气候变化的投入，有利于带动企业和社会的资金投入，为开展应对气候变化的各项工作提供坚实的保障。因而，财政政策为政府减缓和适应气候变化活动提供物质保障，有必要通过优化财政支出结构，增加财政预算用于应对气候变化的支出，逐步建立起稳定投入机制。

（3）财政政策对应对气候变化具有重要的引导作用

财政政策在应对气候变化中的引导作用，主要体现为对节能、能源替代与发展新能源等方面的激励功能和约束功能。财政政策在应对气候变化上的激励作用体现在三个方面。一是鼓励正外部性，即政府对经济主体有利于资源节约和环境保护的正外部性给予其相应的税收减免或财政补贴，把外部效益转化为经济主体的内部效益。二是通过财政补贴、加速折旧、投资抵免等税收支出政策，如对节能电子产品的补贴，加大对环保产业和节能等方面的支持，发挥财政资金引导、鼓励和吸引社会资本投入，形成稳定的环保资金投入渠道。三是通过财政投资、税收优惠等措施，鼓励节能技术、环保新技术的开发和推广。约束功能主要体现在对资源浪费、高耗能产业等的限制和惩罚方面，把产生的负外部性内部化，以提高资源、能源的有效利用，减缓温室气体排放的压力，实现经济与环境的可持续发展。现行税制结构中的增值税、资源税、消费税等，虽并非真正意义上的环境税，但对资源利用、环境保护行为均起到了积极的引导作用和约束功能。如消费税将一部分高污染高耗能产品纳入征税范围，这既有利于筹集资金，也有利于加强环境保护和资源节约①。

（三）省级层面气候公共支出与制度评估的重要意义

1. 有助于有效刻画中国生态文明建设的生动实践

生态文明是中国独特的概念，相较于传统的利用自然、征服自然的思想，生态文明更强调人与自然的相互包涵、和谐共处。中国的生态文明建设实践，就是要探索出符合中国传统思想的现代经济社会发展之路，希望能以中国理念、中国

---

① 财政部财政科学研究所课题组（苏明等）：《中国气候公共财政统计分析研究》，2015年3月。

实践为其他国家提供帮助。地方政府是中国生态文明建设的主要实施主体，对省级层面气候公共支出与制度进行评估，有助于有效刻画中国在生态文明建设方面的生动实践，探索出一条独特的生态文明发展之路。

2. 有利于支持应对气候变化决策

中国当前对于气候财政的统计分析较少，创新性地开展这一工作有利于帮助政府和社会了解气候财政的支出细节、完善预算使用制度、提高资金使用绩效。同时，地方政府开展的应对气候变化工作是在中央政府指导下的地方实验，对省级政府的气候公共支出与制度进行评估，有利于总结各地经验，推进气候变化决策科学化。

3. 有利于中国地方与国外气候财政支出进行比较及交流

全面梳理地方政府气候公共支出信息，科学评估中国地方气候公共支出，有利于中国与其他国家相关气候财政支出进行对比分析，客观评价中国在应对气候变化、生态文明建设方面的成就，同时促进国际交流与合作。

二、河北省气候变化与公共政策及管理现状

河北省是全国资源环境与发展矛盾较为突出的地区之一，环境治理面临的压力与挑战极大，近年来河北省在应对气候变化方面做出了卓有成效的贡献。选取河北作为中国地方政府应对气候变化公共支出分析与评估的案例进行分析，能够以点带面，把握地方政府整体支出情况，也可提取优秀经验，促进地方政府有效应对气候变化。

（一）河北省的气候变化现状

1. 河北地理气候概况

河北省地处华北，北依燕山，南望黄河，西靠太行，东坦沃野，内守京津，外环渤海，周边分别与内蒙古、辽宁、山西、河南、山东等省毗邻。海岸线长487公里，总面积达18.88万平方千米，常住人口7185万，现有170个县级行政区划单位。全省地势由西北向东南倾斜，西北部为山区、丘陵和高原，其间分布有盆地和谷地，中部和东南部为广阔的平原。其中坝上高原平均海拔1200—1500米，占全省总面积的8.5%，燕山和太行山地，其中包括丘陵和盆地，海拔多在2000米以下，占全省总面积的48.1%，河北平原是华北大平原的一部分，

海拔多在50米以下，占全省总面积的43.4%。河北属温带大陆性季风气候，年日照时数2303.1小时，年无霜期81—204天，年均降水量484.5毫米；1月平均气温在3℃以下，7月平均气温18℃至27℃，四季分明①。

据河北省政府2008年发布的《河北应对气候变化实施方案》，河北省近50年来平均气温升高近1.4℃，降雨量减少约120毫米，干旱面积呈扩大趋势，速度为每10年增加1.4%。气候变暖趋势未来将进一步加剧，到2030年河北年平均气温将升高1℃以上；年降水量将普遍增加3%—13%。到2050年，河北年平均气温将升高2.0℃；年降水量增加3%—15%。

2. 河北经济及排放结构

河北省是重工业大省，2015年GDP达到29806.1亿元，第二产业占比48.3%，较全国均值高7.8个百分点，第三产业占比40.2%，较全国均值低10.3个百分点。

河北省的主要产品包括钢铁、机电。根据2014年《世界钢铁统计数据》，2013年全球粗钢总产量为16.06亿吨，中国是世界第一产钢大国，粗钢产量占全球粗钢产量的48.5%。而作为中国第一产钢大省，河北省粗钢产量占到了全球粗钢产量的11.6%，远远超过了世界第二产钢大国日本，比欧盟27国总的粗钢产量还要高。

河北省气候变化治理面临较大挑战，也取得了明显成效。据《中国统计年鉴》，2010年河北工业废气排放总量56324亿标立方米，占该年度全国排放的10.85%，是全国各省份平均排放的3倍余。2014年压减的钢铁产能，占全国压减任务的56%。2015年，河北省各耗能、排放指标均出现下降：单位工业增加值能耗1.64吨标准煤/万元，比2010年下降33.6%；2014年，工业主要污染物COD排放为15.14万吨，氨氮化物为91.8万吨，分别比2010年降低20.7%和21.8%；工业废气主要污染物（二氧化硫、氮氧化物、烟粉尘）占全国排放总量的比重也出现了不同程度的下降；各项产能压减任务也顺利完成。

（二）规划与法规制定

为有序推进应对气候变化工作开展，河北省省委省政府制定了一系列的规划、方案、文件等。国务院2014年发布了《国家应对气候变化规划（2014—2020

---

① 河北概况，数据来源于河北省政府网站。http://www.hebei.gov.cn/hebei/10731222/10751792/index.html。

年)》,其中提到了应对气候变化的 9 类 40 小类措施,本报告按照这一框架,将河北省发布的相关文件(不完全统计)进行了分类(见表 1)。

表 1　河北省近年发布的应对气候变化相关文件

措施		相关文件
控制温室气体排放	调整产业结构 优化能源结构 加强能源节约 增加森林及生态系统碳汇 控制工业领域排放 控制城乡建设领域排放 控制交通领域排放 控制农业、商业和废弃物处理领域排放 倡导低碳生活	河北省装备制造业发展"十三五"规划 河北省工业转型升级"十三五"规划 河北省战略性新兴产业发展"十三五"规划 河北省石化产业发展"十三五"规划 河北省人民政府办公厅关于促进生物产业加快发展的实施意见 河北省节能减排综合性实施方案 河北省人民政府关于实施绿色河北攻坚工程的意见 河北省人民政府关于加快山水林田湖生态修复的实施意见 河北省气象事业发展"十三五"规划 河北省人民政府关于加快推进工业企业技术改造工作的实施意见 《工业和民用燃料煤》(DB13/2081 – 2014) 《洁净颗粒型煤》(DB13/2122 – 2014) 河北省住房城乡建设事业"十三五"规划纲要
适应气候变化影响	提高城乡基础设施适应能力 加强水资源管理和设施建设 提高农业与林业适应能力 提高海洋和海岸带适应能力 提高生态脆弱地区适应能力 提高人群健康领域适应能力 加强防灾减灾体系建设	河北省水利发展"十三五"规划 河北省水权确权登记办法 河北省 2015 年度地下水超采综合治理试点方案 河北省地下水管理条例 河北省保障水安全实施纲要 河北省现代农业发展"十三五"规划 河北省林业"十三五"发展规划 河北省人民政府关于实施绿色河北攻坚工程的意见 河北省海洋环境保护规划(2016—2020 年) 河北省海洋经济发展"十三五"规划 河北省湿地保护规划(2015—2030 年) 河北省"十三五"卫生与健康规划 河北省重污染天气应急预案 河北省大气污染防治条例 河北省突发环境事件应急预案

续表

	措施	相关文件
实施试点示范工程	深化低碳省区和城市试点 开展低碳园区、商业和社区试点 实施减碳示范工程 实施适应气候变化试点工程	开展了低碳试点示范、循环经济示范工作 河北省新型城镇化与城乡统筹示范区建设规划（2016—2020年） 关于开展清洁生产试点示范园区创建工作的意见 河北省大气污染防治计划实施方案 京津冀及周边地区落实大气污染防治行动计划实施细则
完善区域应对气候变化政策	城市化地区 农产品主产区 重点生态功能区	河北省高标准农田建设总体规划（2015—2020年） 河北省建设京津冀生态环境支撑区规划（2016—2020年）
健全激励约束机制	健全法规标准 建立碳交易制度 建立碳排放认证制度 完善财税和价格政策 完善投融资政策	河北省环境污染举报奖励办法 河北省大气污染防治专项资金管理暂行办法（试行） 河北省地质环境恢复治理与保护项目预算定额标准 河北省省级环境保护资金管理使用办法 河北省节能技术改造财政奖励资金管理办法 河北省省级自然保护区专项资金使用管理办法（试行） 河北省工业企业技术改造专项资金管理办法
强化科技支撑	加强基础研究 加大技术研发力度 加快推广应用	河北省节能技术改造财政奖励资金管理办法 河北省工业企业技术改造专项资金管理办法
加强能力建设	健全温室气体统计核算体系 加强队伍建设 加强教育培训和舆论引导	组织开展了应对气候变化统计报表工作 河北省人力资源和社会保障事业发展"十三五"规划 河北省环境污染举报奖励办法
深化国际交流与合作	推动建立公平合理的国际气候制度 加强与国际组织、发达国家合作 大力开展南南合作	
组织实施	加强组织领导 强化统筹协调 建立评价考核机制	河北省应对气候变化实施方案 中共河北省委河北省人民政府关于加快推进生态文明建设的实施意见 河北省生态文明体制改革实施方案

资料来源：根据河北省政府网站资料整理。

如表1所示，河北省近年发布了近50份文件以应对气候变化。这些文件基本与《国家应对气候变化规划（2014—2020年）》的要求相符合，除在国际合作方面缺乏相应措施外，其余领域都发布了相关规划或办法以指导实践工作。这其中，控制温室气体排放与适应气候变化影响领域的规划文件最多，是河北省的工作重点。

在财政支出领域，也明确发布了相关文件，本文共收集到了《河北省环境污染举报奖励办法》等6个文本，相关内容列举如表2所示。

表2　　　河北省应对气候变化相关公共支出安排（部分列举）

文件名称	主要内容
河北省环境污染举报奖励办法	根据举报人对举报事项的举证、协查程度和被举报者对环境危害程度或者因举报避免了重大环境污染损害的，根据环境污染举报案件的性质、内容，确定对举报人的奖金数额，奖励范围为500—3000元。
河北省大气污染防治专项资金管理暂行办法（试行）	专项资金分配采用项目和因素相结合的方式。有具体项目和补助标准的事项严格按照相关标准分配；其他资金按照因素分配。分配因素主要考虑年度污染物减排量、污染治理投入、细颗粒物浓度下降率和上年度考核结果，并结合《河北省大气污染防治行动计划实施方案》规定的重点任务来确定。
河北省省级环境保护资金管理使用办法	环保资金优先支持国家及河北省确定的污染治理和生态保护重点任务，倾斜投向环境污染严重、与人民群众生活密切相关的生态环境领域或项目。 对下补助环保资金采取因素法为主、项目法为辅的分配方式。其中，对按任务量预分配的环保资金，按年度目标实际完成情况据实清算。 申请使用环保资金的省有关部门，可根据本部门所承担的环境保护任务，向省环境保护厅提出资金需求计划并拟定绩效目标。省财政厅、省环保厅综合考虑相关环保任务，统筹安排环保资金。 各市财政、环保部门在接到对下补助环保资金后，要及时提出环保资金细化分配方案，按规定程序批准后，在30日内批复下达，同时报送省财政厅、省环境保护厅备案。 各市财政、环保部门以及使用环保资金的省有关部门应按项目组织开展环保资金绩效评价，形成专题评价报告报送省财政厅、省环境保护厅备案。省财政厅、省环境保护厅在各地各部门绩效评价的基础上有计划地组织开展重点绩效评价。
河北省节能技术改造财政奖励资金管理办法	为了保证节能技术改造项目的实际节能效果，节能资金采取奖励方式，实行资金量与项目节能量挂钩，对企业实施节能技术改造项目给予奖励。 奖励资金支持对象是重点用能企业对现有生产工艺和设备实施节能技术改造的项目；奖励资金主要是对企业节能技术改造项目给予支持，奖励金额按项目实际节能量确定；节能量核定采取企业报告，政府确认的方式。企业提交改造前用能状况、改造后节能措施、节能量及计量检测方法，由政府依据专业机构审核结果予以认定。

续表

文件名称	主要内容
河北省节能技术改造财政奖励资金管理办法	资金的绩效指标由省发展改革委根据项目总体目标，设置衡量项目的绩效指标，确定绩效指标目标值，报省财政厅审核。绩效指标主要包括资金管理指标、产出指标和效果指标，其中：资产管理指标包括资金管理规范性、资金到位情况等指标；产出指标包括项目实施数量、投资规模、节能量等指标；效果指标包括经济效益和社会效益等指标。
河北省省级自然保护区专项资金使用管理办法（试行）	专项资金主要用于：（1）野外自然综合考察、保护区发展与建设规划编制；（2）自然保护区的科研、观察、监测预警仪器设备购置；（3）珍稀濒危物种和生物多样性保护管护设施建设及科研试验；（4）自然生态保护宣传；（5）其他有利于自然保护区事业发展的事项。 自然保护区专项资金应根据绩效目标设置绩效指标，主要包括申报指标、资金管理指标、产出指标和效益指标。其中，申报指标包括申报材料完整性和保护区管理机构独立健全性；资金管理指标包括专项资金使用管理规范性、资金到位情况等；产出指标包括管护设施数量和类型、仪器设备数和类型、规划编制数量和质量、科研实验数量和质量、宣传资料；效益指标包括建设项目运行效果、野生动植物保护效果、生态环境保护效果。
河北省工业企业技术改造专项资金管理办法	专项资金以支持我省年度"千项技术改造工程①"项目为主，重点向"十百千"工程、"三个一百"工程、龙头带动计划以及对标示范企业倾斜。 专项资金综合采取事后补贴、贷款贴息、股权投资和购买服务等支持方式。事后补贴，即对全省工业转型升级有示范带动作用的重大技改项目，根据设备购置等固定资产投入资金额度给予一定比例补助，单个项目最高不超过2000万元；对公共平台项目的固定资产投资给予不超过20%的补助，单个项目最高不超过500万元。贴息按照项目贷款实际发生额和人民银行公布的同期贷款基准利率计算，支持额度不超过12个月的利息额，单个项目最高不超过2000万元。 资金的绩效指标由省工业和信息化厅根据项目的总体目标，设置衡量项目绩效指标。绩效指标主要包括资金管理指标、产出指标和效果指标，其中：资金管理指标包括资金管理规范性、资金到位情况等；产出指标包括技改项目实施数量等指标；效果指标包括四新成果（新技术、新工艺、新装备、新材料）利用率等经济效益指标。

资料来源：根据河北省财政厅网站资料整理。

---

① "千项技术改造项目"为河北省2014年设立的、每年滚动的省工业企业重点技术改造项目，这些项目已开工或基本具备开工条件，项目的实施对于调整河北省产业结构、转变发展方式，稳定经济增长将有重大意义。其后的"十百千工程"是省政府提出的推进工业转型升级、打造工业强省的实施措施，具体包括壮大十大工业基地、扶持百家优势企业、培育千项名牌产品。"三个一百"工程指的是河北省为努力扩大有效投资，要重点抓好的100个续建保投产、100个新开工和100个前期项目。

由于缺乏相应的财政统计分析，我们依据调研了解的情况和公开可获取的资料梳理了河北省应对气候变化相关的财政支出政策（见附件（一））。

（三）河北省气候变化管理机构

河北省委、省政府高度重视气候变化问题，于2008年2月成立了"河北省应对气候变化领导小组"，省长任组长，建立了应对协调机制，并根据可持续发展战略的要求，制定《河北省应对气候变化实施方案》，采取了一系列与应对气候变化相关的政策和措施，为减缓和适应气候变化作出了积极努力（见图1、图2）。河北省发展改革委员会下设应对气候变化处，承担省应对气候变化领导小组办公室日常工作。河北省气候相关政府部门及其职责（见附件（二））。

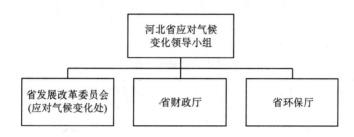

图1 河北省应对气候变化宏观协调机制

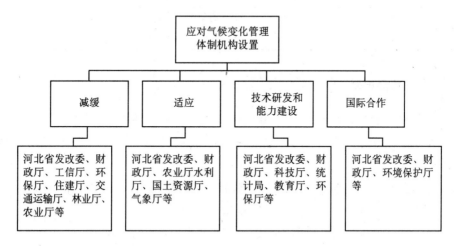

图2 河北省应对气候变化管理体制机构设置

三、河北省气候公共支出预算制度

（一）预算管理层级

中国是一个政体统一的国家，由各级政府履行职能。财政为各级政府职能的实现提供财力保证。因此，有一级政府即有一级财政收支活动主体，也就应有一级预算管理主体。《中华人民共和国预算法》第3条规定，国家实行一级政府一级预算，设立中央，省、自治区、直辖市，设区的市、自治州，县、自治区、不设区的市、市辖区，乡、民族乡、镇五级预算（见表3）。河北省预算层级包括省、市、县、乡四级。

表3　　　　　　　　　　政府预算管理主体

层级	预算管理政府
一级	中央
二级	省、自治区、直辖市
三级	设区的市、自治州
四级	县、自治区、不设区的市、市辖区
五级	乡、民族乡、镇

（二）预算编制

部门预算是各级政府总预算编制的基础。按照《中华人民共和国预算法》的有关规定，政府预算的编制一般采用自上而下、自下而上、上下结合、逐级汇总的程序（见图3）。

省、自治区、直辖市政府根据国务院的指示和财政部的部署，结合本地区的具体情况，提出编制本行政区域预算草案的要求。

县级以上地方各级政府财政部门审核本级各部门的预算草案，编制本级政府预算草案，汇编本级总预算草案，经本级政府审定后，按照规定期限报上一级省、自治区、直辖市政府财政部门、财政部门汇总的本级总预算草案，应当于下一年1月10日前报财政部。

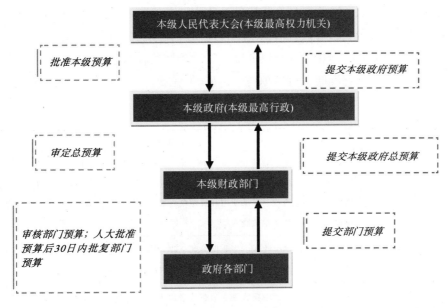

图3 预算编制流程图

地方各级政府预算草案经本级人民代表大会批准本级政府预算。县级以上地方各级政府财政部门应当自本级人民代表大会批准本级政府预算之日起30日内，批复本级各部门预算。地方各部门应当自本级财政部门批复本部门预算之日起15日内，批复所属各单位预算。

（三）预算科目设置

为加强收入管理和数据统计分析，根据中国政府收入构成情况，结合国际通行分类方法，财政预算收入按经济性质分为类、款、项、目；支出按功能分类分为类①、款②、项③，按其经济性质分为类、款。各级预算支出应当按其功能和经济性质分类编制。

1. 收入分类

按照全面、规范、细致地反映政府各项收入来源和性质的要求，政府收入分为"类、款、项、目"四级，四级科目逐级细化，以满足不同层次的管理需要。

---

① 综合反映政府职能活动，如国防、外交、教育、科学技术、社会保障和就业、环境保护等。
② 反映为完成某项政府职能所进行的某一方面的工作，如"教育"类下的"普通教育"。
③ 反映为完成某一方面的工作所发生的具体支出事项，如"水利"款下的"抗旱"、"水土保持"等。

政府收入划分为税收收入、社会保险基金收入、非税收入、贷款转贷回收本金收入、债务收入以及转移性收入等（见表4）。其中税收收入分设21款，社会保险基金收入分设6款，非税收入分设8款，贷款转贷回收本金收入分设4款，债务收入分设2款，转移性收入分设10款。

表4　　　　　　　　　　　预算收入科目设置

类	款
税收收入	增值税、消费税、营业税、企业所得税、企业所得税退税、个人所得税、资源税、固定资产投资方向调节税、城市维护建设税、房产税、印花税、城镇土地使用税、土地增值税、车船使用和牌照税、船舶吨税、车辆购置税、关税、耕地占用税、契税、其他税收收入
社会保险基金收入	基本养老保险基金收入、失业保险基金收入、基本医疗保险基金收入、工伤保险基金收入、生育保险基金收入、其他社会保险基金收入
非税收入	政府性基金收入、专项收入、彩票资金收入、行政事业性收费收入、罚没收入、国有资本经营收入、国有资源（资产）有偿使用收入、其他收入
贷款转贷回收本金收入	国内贷款回收本金收入、国外贷款回收本金收入、国内转贷回收本金收入、国外转贷回收本金收入
债务收入	国内债务收入、国外债务收入
转移性收入	返还性收入、财力性转移支付收入、专项转移支付收入、政府性基金转移收入、彩票公益金转移收入、预算外转移收入、上年结余收入、调入资金

2. 支出功能分类

支出功能分类主要反映政府职能活动的不同功能和政策目标，说明政府究竟做了什么事。政府支出功能分类分为"类、款、项"三级科目。类级科目综合反映政府职能活动；款级科目反映为完成某项政府职能所进行的某一方面的工作；项级科目反映为完成某一方面的工作所发生的具体支出事项。三级科目由大到小、由粗到细，分层次设置。一般公共预算支出按照其功能分类，分别为17类、170多款、800多项。包括一般公共服务支出，外交、公共安全、国防支出，农业、环境保护支出，教育、科技、文化、卫生、体育支出，社会保障及就业支出和其他支出。

四、河北省气候公共支出统计与评估

（一）河北省气候公共财政统计范围的确定

1. 减缓气候变化的支出

——能源结构调整方面：可再生能源支持、清洁煤技术支持、核电支持等。

——产业结构调整方面：淘汰钢铁落后产能；发展战略性新兴产业专项资金；支持传统产业技术升级和结构调整项目。

——节能和提高能效方面：促进节能降耗、资源综合利用和生态建设的支出。

——发展排污权交易方面：全省主要污染物排放权交易的技术性、事务性工作，全省排污权交易网络及平台建设、管理及维护工作，为主要污染物排放权交易活动提供相关服务。

2. 适应气候变化的支出

——农业方面：农业资源保护修复与利用、防灾救灾、草原植被恢复费安排的支出等方面的支出。

——水利方面：水资源节约管理与保护、水利工程建设等方面的支出。

——气象方面：气候变化监测、预估、评估方面的支出。

——城市基础设施建设方面：城镇污水处理设施和管网配套建设、城市垃圾无害化处理和配套设施建设、建筑节能材料产品、城市规划区的绿化工作等的支出。

3. 气候能力建设的支出

——机构和人员方面：政府相关气候部门和机构运转及支出，河北省环境保护资金项目专家库建设支出。

——技术研发支出。

——环境保护宣传教育支出。

（二）河北省气候公共支出统计

1. 基本研究思路

关于气候统计的具体做法，参考中国气候公共财政统计分析研究项目第一阶段的方法，本报告根据河北省的现实情况，采取以政府收支分类为主、同时根据具体的支出项目分项目甄别的做法。首先根据政府收支分类科目定位涉及气候的项目，然后再结合各个科目下的支出项目，按照与气候的相关程度进行细分。

2. 省级政府收支分类与气候相关程度的划分

在参考联合国开发计划署（UNDP）主持研究并提供的亚太地区多国气候支出分析分类方法的基础上，结合河北省目前的政府收支分类中类、款、项的设置，本报告按照与气候的相关程度划分财政支出（见表5）。

表5　　　　省级一般公共预算支出分类与气候的相关度（简表）

序号	科目名称	高度相关	中度相关	低度相关	不相关
1	一般公共服务支出		√	√	√
2	国防支出		√	√	√
3	公共安全支出				√
4	教育支出				√
5	科学技术支出		√	√	
6	文化体育与传媒支出				√
7	社会保障和就业支出		√		
8	医疗卫生支出	√	√	√	
9	节能环保支出	√	√		
10	城乡社区支出			√	√
11	农林水支出	√	√		
12	交通运输支出				√
13	资源勘探信息等支出			√	
14	商业服务业等支出				√
15	金融支出				√
16	援助其他地区支出				
17	国土海洋气象等支出	√	√		
18	住房保障支出				√
19	粮油物资储备支出				√
20	预备费				√
21	其他支出				√

（1）高度相关：支出明确的首要目标与应对气候变化直接相关，包括节能环保中的环境监测与监察、污染防治、自然生态保护、天然林保护、退耕还林、风沙荒漠治理、退牧还草、能源节约利用、污染减排、可再生能源、资源综合利用，社会保障中的自然灾害生活救助，公共卫生中的疾病预防控制、突发公共卫

生事件应急处理、基本公共卫生服务、重大公共卫生专项，农业中的灾害救助、农业资源保护与利用，林业、防灾减灾等支出。

（2）中度相关：支出的次级目标或部分目标与应对气候变化相关，包括农林水中的水利、南水北调支出，节能环保中的能源管理事务、一般公共服务中的发展与改革事务、环境保护管理事务，科技支出中的基础研究，公共卫生中的应急救治机构、其他专业公共卫生机构、妇幼保健机构支出，城乡社区中的国家重点风景区规划与保护、城乡社区规划与管理、城乡社区公共设施支出，交通运输中的铁路等支出。

（3）低度相关：支出中包括或涉及与应对气候变化相关的支出，如一般公共服务中的财政事务及统计信息事务，科学技术支出中的科学技术管理事务、应用研究、科技条件与服务、科学技术普及，城乡社区支出中的城乡社区管理事务，农林水支出中的农业行政运行、扶贫支出，交通运输中的公路水路运输，公检法等公共安全支出等。

（4）不相关：支出与应对气候变化不直接相关，如一般公共服务中的人大事务、政协事务、人力资源事务等，国防支出等，或无法标识的支出，如教育支出、文化体育与传媒支出。

由表5看出，节能环保支出、农林水支出与气候相关度较高，以此两类为例，本报告对不同层级政府投入进行分析。

由图4和图5看出，在节能环保和农林水支出中，县级政府投入最多，其次是地级政府，乡镇政府与省级政府在投入中所占比重较小。

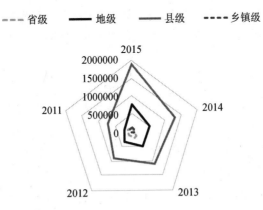

图4　2011—2015年度河北省各级政府节能环保支出图（单位：万元）

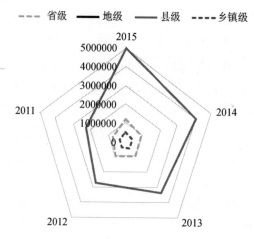

图 5　2011—2015 年度河北省各级政府农林水支出图（单位：万元）

根据气候相关度本报告对河北省一般公共预算支出进行分类，列表反映了与气候高度相关、中度相关、低度相关的公共支出分类科目（见附件（四））。

3. 统计结果及分析

根据气候支出的相关度分析，河北省 2011—2015 年全省财政支出中，气候高度相关支出占比分别为 4.78%、4.69%、5.49%、6.02%、6.93%，气候中度相关支出占比分别为 3.44%、4.06%、4.02%、4.44%、4.35%，高度相关与中度相关支出合计占比分别为 8.23%、8.75%、9.51%、10.46%、11.29%（见图 8）。测算数据表明，河北省高度重视气候变化问题，"十二五"期间，气候公共支出逐年增长。从气候公共支出在全省财政支出的占比来看，河北省的气候公共支出占比高于中央气候相关支出的同口径数据。根据项目第一期的数据测算，2013—2014 年，中央层面高度相关与中度相关气候支出合计占财政支出的比例为 7.6%、6.9%[①]。

具体看，与气候高度相关的财政支出整体呈上升趋势，与气候中度相关的支出从 2011 年的占比 3.44% 上升到 2015 年的 4.35%，与气候低度相关的决算支出呈总体下降趋势（见图 6）。

按照上述测算的气候相关支出比例，可得各年度气候相关支出情况。测算得出，2011—2015 年河北省财政支出中，与气候变化高度相关支出分别为 169.20 亿元、191.45 亿元、242.25 亿元、281.55 亿元、390.50 亿元。数据显示，"十二五"期间，河北省气候相关支出逐年增长。尤其是 2015 年河北省与气候变化

---

① 财政部财政科学研究所课题组（苏明等），中国气候公共财政统计分析研究，2015 年 3 月。

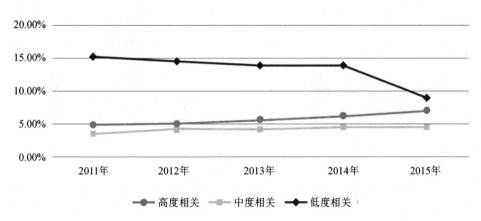

图 6　按气候相关度划分的河北省全省财政支出比重变动图

高度相关支出较上年增长达 38.7%，这主要与河北省 2015 年以来加大节能环保等方面的投入有关（见图 7、图 8）。

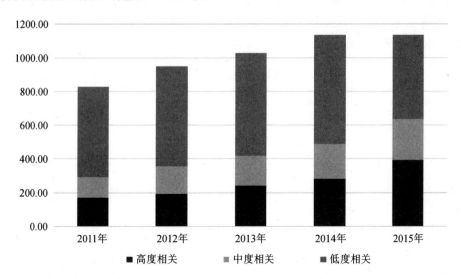

图 7　河北省 2011—2015 年气候相关公共支出图

由图 7 看出，河北省一般公共支出中，与气候相关的财政支出的绝对数呈上升的趋势。由图 8 看出，与气候变化高度相关、中度相关的财政支出占河北省当年 GDP 的比重呈上升趋势，与气候变化低度相关的财政支出呈下降趋势。产生这种现象的原因是河北省合理调整与气候相关的财政支出，将更多的财政资金投入到与气候变化相关度更高的领域，提高财政资金使用的针对性和使用效率。综上可得，河北省无论是在气候财政投入的绝对数量，还是气候财政投入占当年

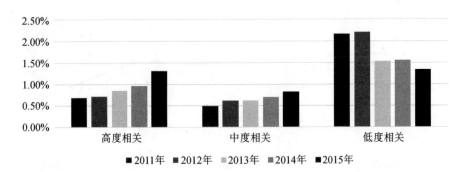

图8 河北省2011—2015年气候相关支出占河北省当年GDP比重

GDP的相对比重，都呈上升趋势，且资金投入更为合理有效。

根据与气候相关程度，结合河北省财政支出的具体情况，对高、低、中度相关性赋予百分比，对河北省气候相关支出进行更为精确的测算。赋值方案见表6。根据各赋值方案测算后的河北省气候公共支出占财政支出的总比例见表7。测算结果显示，2015年，河北省气候公共支出占财政支出的比重最乐观估计为12.09%，最保守估计为8.44%。

表6 各气候相关度赋值方案[①]

赋值方案	高度相关	中度相关	低度相关
方案1	100%	70%	30%
方案2	100%	50%	20%
方案3	80%	50%	20%
方案4	70%	50%	20%
方案5	90%	50%	20%

4. 对现有分析方法的进一步反思

我们认为，现有分析方法存在进一步改善的空间，相关度判断的原则还需要调整。例如，某项活动按照减缓气候变化的标准与按照适应气候变化的标准进行

[①] 以上百分比带有一定的模拟性。由于从公开路径无法对与气候相关的财政支出进行进一步的细分和统计，在借鉴相关国家的统计经验的基础上，根据气候财政相关度的差异，对气候支出的比重进行预判。如方案1是指，假设高度相关的财政支出中100%的财政支出用于应对气候变化，中度相关的财政支出中有70%的财政支出用于应对气候变化，低度相关的财政支出中有30%的财政支出用于应对气候变化。其他方案以此类推。

表7　　　　　　不同情景气候相关支出占财政支出比例分析

赋值方案	2011	2012	2013	2014	2015
方案1	11.69%	11.85%	11.26%	12.08%	12.09%
方案2	9.50%	9.60%	9.47%	10.21%	10.52%
方案3	8.55%	8.66%	8.37%	9.01%	9.13%
方案4	8.07%	8.20%	7.82%	8.40%	8.44%
方案5	9.03%	9.13%	8.92%	9.61%	9.82%

气候相关性的判定，得出的结果会有所差异，进而进一步影响相关财政资金的统计。

应对气候变化的两个主要方面是缓解（削减温室气体的排放）和适应（承认变化并建立各种制度加强我们的应变能力）。无论是发达国家还是发展中国家，都必须采取"可测量，可报告和可核实的"减排行动。减排措施旨在处理气候变化的原因，适应气候变化的重点则是处理气候变化的影响。适应是指采取政策和做法来应对气候变化的影响，各部门都有种种适应的办法，例如关于水资源，可以扩大雨水收集、蓄水、节约水；在农业方面，调整种植日期和作物品种，作物搬迁；在基础设施（包括沿海地区）建设上，建立湿地，作为对海平面上升和洪水的缓冲；对能源使用方面，使用可再生能源，提高能源效率。

根据减缓气候变化与适应气候变化的界定，结合中国河北省的实践，本报告对预算支出表科目的气候相关性进行重新判定，具体结果见附件（五）表1—表3。

按照减缓气候变化与适应气候变化的界定，对预算支出表科目的气候相关性进行重新判定后发现：在高度相关中，减缓气候变化科目数量占高度相关科目数量总和的36%，适应气候变化科目数量占高度相关科目数量总和的25%，而两类活动都高度相关的科目占高度相关科目数量总和的39%；在中度相关中，减缓气候变化科目数量占中度相关科目数量总和的19%，适应气候变化科目数量占中度相关科目数量总和的42%，而两类活动都中度相关的科目占中度相关科目数量总和的38%；在低度相关中，减缓气候变化科目数量占低度相关科目数量总和的20%，适应气候变化科目数量占低度相关科目数量总和的36%，而两类活动都中度相关的科目占低度相关科目数量总和的44%（见图9）。

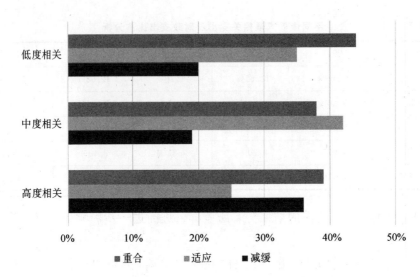

图9 一般公共预算支出科目按气候相关度重分类结果分析图

以河北省2015年一般公共决算支出为例,按照重新划分的气候相关性标准,进行重新统计计算。发现在高度相关性中,减缓与适应重合的金额占高度相关决算支出的21.19%;在中度相关性中,减缓与适应重合的金额占中度相关决算支出的20.02%;在低度相关性中,减缓与适应重合的金额占低度相关决算支出的56.06%(见表8)。分析具体决算科目发现,其重合之处大部分为机构运行类支出、科研类支出、宣传类支出等,此类活动同时具有减缓与适应的活动属性。按照重新分类统计的气候公共支出占当年一般公共决算支出及GDP的比重见表9和表10。

表8 2015年河北省气候相关支出重新统计结果表　　单位:亿元

	减缓	适应	减缓和适应	重合百分比(%)
高度相关	336.61	136.67	82.73	21.19%
中度相关	209.5	84.85	49.11	20.02%
低度相关	445.24	335.2	280.44	56.06%

表9 重新统计结果占当年河北省财政一般公共决算支出的比重

	减缓	适应	减缓和适应
高度相关	5.98%	2.43%	1.47%
中度相关	3.72%	1.51%	0.87%
低度相关	7.91%	5.95%	4.98%

表 10　　　　　　　重新统计结果占当年河北省 GDP 的比重

	减缓	适应	减缓和适应
高度相关	1.13%	0.46%	0.28%
中度相关	0.70%	0.28%	0.16%
低度相关	1.49%	1.12%	0.94%

由以上统计分析可看出，在高度相关、中度相关、低度相关的预算支出科目中，有超过半数以上的支出科目可以根据减缓气候变化与适应气候变化的界定，对科目进行二次细分。进行二次细分后，可更精确地统计用于应对气候变化的财政支出数额。

五、主要结论与建议

（一）主要结论

1. 气候变化问题日益受到重视，气候公共支出逐年增长

自十八大以来，中国始终把生态文明建设放在治国理政的重要战略位置。三中全会提出加快建立系统完整的生态文明制度体系，四中全会要求用严格的法律制度保护生态环境，五中全会将绿色发展纳入新发展理念。河北省高度重视气候变化问题。对河北省气候公共支出评估结果显示，"十二五"期间，气候公共支出逐年增长。尤其是 2015 年以来河北省加大节能环保等方面的投入，与气候变化高度相关支出较上年增长达 38.7%。根据气候支出的相关度分析，2011—2015 年，与气候高度、中度相关的财政支出整体呈上升趋势，全省财政支出中与气候变化高度相关与中度相关支出合计占比从 8.23% 上升到 11.29%。从气候公共支出在全省财政支出的占比来看，河北省的气候公共支出占比高于中央气候相关支出的同口径数据。

2. 气候变化资金投入来源多渠道，投入方式多元化，投入方向立体化

对河北省气候公共支出与制度评估表明，河北省在应对气候变化过程中，资金投入来源多渠道，形成包括政府、资本市场、国际贷款的多渠道投入。其中，政府财政投入除却一般公共预算支出外，还有国有资本支出、政府性基金支出等。在资金投入总量上，中国各级政府在气候变化领域的财政投入呈上升趋势。

在资金投入方式上，除了以专项资金、以奖代补等方式的财政支出外，探索运用市场化方式（排放权交易）和金融工具（发行绿色证券），实现对新型产业、环保产业、低碳产业的支持。在资金投入方向上，覆盖减缓与适应气候变化两方面。在减缓气候变化中，初步从末端治理延伸到源头防控，对排放物的处理实现从天空到土壤的防治；在适应气候变化中，涉及管理机构的设置与职能调整、人体健康、生活方式、出行方式、生产方式的倡导与改变等多方面。

3. 气候公共支出预算制度更加完善，为资金使用效率提高和加强资金监管提供制度保障

政府预算体现政府活动的范围以及政府在特定时期所要实现的政策，预算制度则是政府实现政策目标的保障。随着预算管理制度改革的不断深化，气候变化公共支出的预算科目不断完善。2015年，政府预算支出科目中增加"节能环保"类，财政预算中增加了专门反映与应对气候变化高度相关的科目；初步建立了全口径预算管理体系，建立政府性基金预算、国有资本经营预算、社会保险基金预算与一般公共预算的统筹衔接机制。强化本级与上级资金统筹，提高资金使用的透明度。改进预算控制方式，通过中期财政规划管理、跨年度预算平衡机制，加强预算与规划的关联衔接，改进财政资金拨付及使用，减少财政资金沉淀，提高资金使用效率。

4. 气候公共支出统计评估方法得以拓展，气候财政的方法研究得以深化

在中国气候公共支出与制度评估项目第一阶段，仅按照与应对气候变化的相关度高低将气候财政资金进行分类统计，评估中国中央层面应对气候变化的公共支出情况；项目第二阶段，本报告对气候公共支出分类统计方法进行了拓展，在按照气候变化相关度对河北省气候公共支出进行分类统计的基础上，又按照减缓气候变化与适应气候变化两个维度，对河北省气候变化财政资金按照相关度进行统计分析。本报告发现，对气候变化公共支出按照减缓气候变化与适应气候变化活动的相关度进行二次细分的统计分析方法，可以更加精确地统计用于应对气候变化的公共支出数额，为公共支出的成本效益分析与绩效评价打下基础。

（二）相关建议

1. 应对气候变化，应强化理念渗透和政策的顶层设计，形成政府与市场的良性互动

在国家治理层面已经多次重申"既要绿水青山，也要金山银山"、"宁要绿水青山，不要金山银山"、"绿水青山就是金山银山"的发展理念，在执行层面

上，尽管存在当地税收减少、新的经济增长点未形成之际经济增长呈现下滑等客观因素，但河北省仍然加强对影响气候变化的消极举措的监管。在采取的应对措施上，地方政府在采用行政手段的同时，更多的是行政手段与市场手段相结合。因此，在政策设计中，要明确理念，注重顶层制度建设，用系统性、全局性的思维指导应对气候变化的措施。政府应与市场形成良好的互动，在尊重市场规律的前提下，更多地通过市场化手段，应对气候变化。通过提高行业准入门槛、税收、收费等举措提高其污染成本，将污染成本内化，倒逼企业摆脱"高投入、高污染、低利用率"的粗放发展模式，走绿色发展道路。此外，利用价格规律，政府通过加大对节能低耗产品的采购，影响市场需求，从而促进生产的转型升级。

2. 继续深化预算管理制度改革，提高政策和资金的有效性

为应对气候变化，河北一方面增加公共资金的投入，另一方面强化预算资金的绩效，进行了许多探索，做出了卓有成效的工作。建议进一步完善预算管理制度。一方面，在财政预算管理中引入中期的理念，实行中期财政规划管理，使预算安排与政府中长期政策相衔接，推动应对气候变化的政府决策由年度决策转向中期决策，更具前瞻性和连续性，确保当前政策与财政的长期可持续性相一致。另一方面，实行全过程预算绩效管理，提高绩效信息质量，实现绩效管理与评价结果与预算安排相结合。在预算编制时编报绩效目标，建立规范的绩效指标体系，量化的绩效指标不仅包括产出的数量、质量、时效、成本指标，还应包括经济效益、社会效益、生态效益、可持续影响等效益指标，以及服务对象满意度指标。开展气候公共支出政策绩效评价，编制政策绩效评价报告，并向社会公开。

3. 进一步创新财政支出方式，建立市场化的约束机制

应对气候变化是一项长期的、艰巨的任务。对气候变化公共支出的评估显示，近年来，河北省应对气候变化的公共支出不断增多，注重创新财政支出方式改革，气候支出资金使用效益不断提高。但是，面对应对气候变化的巨量资金需求，单靠政府财政一己之力恐是捉襟见肘。除了拓宽资金投入渠道外，需要进一步创新财政支出方式，加强财政与金融工具（基金、债券等）的融合，提高财政资金的导向作用，发挥"四两拨千斤"的作用。例如通过PPP模式，通过财政贴息，绿色金融方式，吸引社会资本进入相关领域。更多地运用排放权交易等市场化方式，运用绿色债券等金融工具，支持新型产业、环保产业、低碳产业的发展。

4. 加强气候公共支出成本效益评估的研究

气候公共支出评估是一个崭新的研究领域，除了需要建立科学的分类统计方法，评估气候公共支出的总体规模和结构之外，还需要对气候公共支出的成本效益进行全面评估，分析气候变化涉及的各相关利益主体的成本，以及在生态环境、社会、经济等各个层面产生的效益，为气候公共支出政策的完善提供依据。

报告二：气候公共支出的成本效益分析
——以河北省"去产能"为例

引言

气候变化问题与经济发展方式密切相关。促进经济发展方式向低碳转型是应对气候变化的必然选择。当前，中国正在推进供给侧结构性改革，去产能是供给侧结构性改革的重要内容。过去长期粗放式发展惯性作用下形成的产能过剩，不仅加大了当前经济下行压力，也在环境保护、资源节约等领域形成了很多短板，不仅资源环境难以承受，发展也将难以为继。实施去产能，能够优化升级生产结构，提升资源利用效率，降低工业排放，提高经济发展质量，是应对气候变化的重要举措。

在应对气候变化方面，河北省具有典型性。河北是工业大省，2016年全省生产总值实现31827.9亿元，第二产业增加值15058.5亿元，占比47.3%，较全国水平高7.5个百分点，工业在河北经济中占有重要地位。河北省是钢铁大省，2016年，生铁产量18398.4万吨，粗钢产量19260.0万吨，钢材26150.4产量万吨。钢铁产量占全国产量的1/4。近年来，河北省积极应对气候变化，不断增加气候相关公共支出，应对气候变化取得了一定成效。

在评估河北省气候公共支出与制度的基础上，本报告选取气候变化公共支出的具体项目进行成本效益分析，意在探索气候公共支出绩效评估的方法框架。鉴于去产能是近年来河北省应对气候变化的重要举措之一，我们选取河北省"去产能"项目进行研究，结合对河北省相关政府部门、企业的实地走访调研，较为系统地分析河北省去产能的成本与效益,探索构建气候公共支出成本收益分析的方法框架。

一、河北省去产能进展

2013年9月,习近平总书记就化解过剩产能发表重要讲话,随后明确要求河北要决战决胜,打好调整经济结构、化解过剩产能这场攻坚战。2013年,河北省制定实施了《化解产能严重过剩矛盾实施方案》,提出化解产能过剩的"6643"工程:计划到2017年压减6000万吨钢铁、6000万吨水泥、4000万吨标煤、3000万标准重量箱平板玻璃产能。各项工作依据该实施方案推行良好,部分行业超量完成任务。

(一)河北去产能行业基本情况

河北是工业大省,相关工业产品产量占全国的比重极高,2012年主要工业产品占比列举如表11所示,除水泥、玻璃产量占比较低外,钢铁产量在全国占比都在20%以上。

表11　　　　　　　　2012年河北省相关工业产品产量及占比

	粗钢（万吨）	钢材（万吨）	生铁（万吨）	水泥（万吨）	玻璃（万重量箱）
全国	72388.2	95577.8	66354.4	220984.1	75050.5
河北	18048.4	20995.2	16350.2	12809.8	11382.7
占比	24.93%	21.97%	24.64%	5.80%	15.17%

资料来源:国家统计局、河北省统计局。

(二)近年河北去产能进展

产量占比大使得河北去产能工作压力较重,例如,2014年河北省在钢、铁、水泥、平板玻璃产业完成的去产能任务,分别占当年度全国计划任务的55.56%、55.56%、93.29%、72.37%[①]。河北省高度重视去产能工作,工作完

---

① 资料来源于中国国务院新闻办公室网站。

成情况较好。

如图 10 所示，虽然略有波动，但河北 2013—2016 年保持了主要产品年均 1000 单位以上的去产能工作成绩。2014 年度的去产能工作完成的最为突出，当年度水泥和平板玻璃产能减少了 3918 万吨和 2533 万重量箱，大大缓解了后续去产能任务的压力，极大推动了产能布局调整。

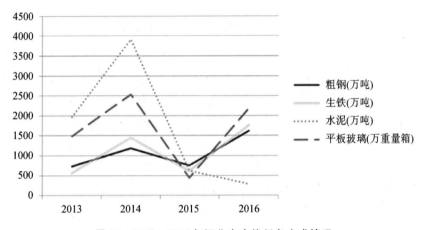

图 10 2013—2016 年河北去产能任务完成情况

"十二五"期间，河北省共压减炼铁产能 3391 万吨、炼钢 4106 万吨、压减水泥产能 13834 万吨、平板玻璃 7202.5 万重量箱①，分别完成原定 6643 计划的 68.43%、56.51%、230.57%、240.08%，水泥和平板玻璃都已超额完成任务。就炼钢和炼铁的产能减产规模而言，分别占全国同期减产规模的 37.26% 和 43.31%。同时在此期间，虽然 2014 年经济增速同比下滑 1.7 个百分点，但近 3 年经济增速保持在 6.8% 左右，与全国水平持平，经济发展较为稳定，河北经济没有出现断崖式下滑，2016 年末城镇登记失业率为 3.68%，低于全国水平 0.34 个百分点，没有发生大规模失业，没有发生区域性系统风险，总体任务进展良好。

（三）未来河北去产能规划

"十三五"期间，河北省还将承担全国 1/3 的钢铁产能压减任务，计划 2016—2017 年压减炼铁产能 3715 万吨、炼钢 3177 万吨，"十三五"期间计划压

---

① http://www.hbdrc.gov.cn/web/web/xwbd/4028818b555274660155d76169f978f6.htm。

减炼铁产能 4989 万吨、炼钢产能 4913 万吨①,到"十三五"末,将钢铁、水泥、平板玻璃产能分别控制在 2 亿吨、2 亿吨、2 亿重量箱左右。

二、去产能的成本分析

"去产能"即化解产能过剩,是指为了解决产品供过于求而引起产品恶性竞争的不利局面,寻求对生产设备及产品转型升级的方法。产能过剩是资源错配的表现,大量宝贵的社会资源消耗在产能过剩的行业,导致资源未得到充分利用。"去产能"不仅是淘汰落后产能,更重要的是产业结构的优化升级。"去产能"旨在通过工业领域生产方式的变革,实现由粗放经济向循环经济的转变,减少工业废弃物对环境的污染与破坏,降低人类活动对气候的负面影响。对河北来说,"去产能"更是生态文明建设的重要组成部分,对改善居民的生活环境、对气候改善有重要意义。为实现"去产能"的目标,各级政府、企业等相关利益主体承担起相应的职责,充分发挥各自的职能,在去产能过程中付出很大的成本代价。

(一)去产能成本分析的理论框架

为了全面地预估、测算"去产能"的成本,首先要明确"去产能"的内涵及外延特征、表现形式、成本范围。在此,根据"去产能"过程中政府、企业采取的行动与措施,合理、全面地进行成本统计与测算。

1. "去产能"成本的多维度认识

在会计学中,成本根据用途的不同会有不同的定义。美国会计学会(AAA)所属的"成本与标准委员会"对成本的定义是,"为了达到特定目的而发生或未发生的价值牺牲,它可用货币单位加以衡量"。中国成本协会(CCA)发布的 CCA2101:2005《成本管理体系术语》标准中第 2.1.2 条中对成本术语的定义是:"为过程增值和结果有效已付出或应付出的资源②代价"。

成本问题在某种意义上说是一个分配问题,是一个相互关联的系统性问题。"去产能"过程中部分成本的支出存在滞后性、潜伏性的特征(例如,去产能导

---

① http://www.hbdrc.gov.cn/web/web/xwbd/4028818b55144edf015542058f042a8c.htm。
② 是指凡是能被人所利用的物质。在一个组织中资源一般包括:人力资源、物力资源、财力资源和信息资源等。

致社会风险及金融风险上升。为维持经济平稳运行，财政需对此产生的风险进行兜底）。若仅从会计角度考虑，则容易出现忽视长期隐性成本，导致成本统计不全面的结果。

本报告认为，对去产能成本研究和测算，必须将成本的定义建立在经济学、管理学和社会学理论基础之上。"去产能"的成本包括看得见的、可计量的资金投入成本，也包括制度层面的转型成本（刘尚希，2016）。将"去产能"放在生产方式转变、经济社会转型和全球化视角中多角度、多维度考察，才有可能得出科学全面的"去产能"成本。由此，本报告定义"去产能"的成本可分为显性成本、隐性成本、机会成本。

"去产能"的显性成本指"去产能"过程中直接的、当下的支出，具有一次性、静态的特点。"一次性"指资金投入形式具有明显的阶段性特征，一次性成本的具体范围相对明确，也易于测算，例如基础设施、职工安置的投入等；"静态"指一定价格下，就某个时点计算出来的成本。

隐性成本指"去产能"过程中非直接和非当下的支出，具有经常性、动态的特点。"经常性"体现在基础设施的运行维护成本，经常性的社会管理成本等，具有明显的刚性支出特征，且其具体范围数额测算等都可能是动态的。"动态"的特点使我们在成本统计时要考虑一段时间内价格变化因素，包括劳动力成本、基本建设材料价格、公共服务成本变化等带来的"去产能"整体成本的变化。从现实来看，中国"去产能"动态成本就有明显的上升特征。

机会成本指"去产能"过程中所要放弃另一些东西的价值，也可以理解为在面临多方案择一决策时，被舍弃的选项中的价值。

"去产能"是系统化、全局化的行动，需要市场主体的配合，也需要政府的适当干预。

2. 政府"去产能"成本

对政府来说，"去产能"的成本包括看得见的、可计量的资金投入成本，也包括制度层面的转型成本。对于政府，"去产能"本身蕴含着发展改革转型的全面要求，实际上它涵盖了生产方式转变，其核心目的是实现资源的优化配置，实现生产方式的转变。所以，"去产能"进程离不开发展理念的转变和各项体制改革的深化，这包括经济发展方式的转变，注重生态环境保护等。

政府"去产能"的显性成本，包括但不仅限于政府为了鼓励落后产能退出，设立奖补资金的支出；为防范企业"去产能"带来的社会风险，增加的安置职工支出；为更好地掌握"去产能"的进度，使"去产能"过程公开透明，政府机关建立项目信息库和公开制度增加的支出;"去产能"过程中,为降低企业生产过程中废弃物排放对生态环境的破坏,保护环境,应对气候变化,对企业排污进行实时监控,尤其是河北省成立全国唯一的环保执法队,增加社会管理成本;重新进行城市规划,引导企业搬迁而发生的产业转移成本,等等。

政府"去产能"的隐性成本包括但不仅限于给予"去产能"企业的税收优惠;由于"去产能"过程中,金融系统收紧对特定行业的信贷活动,因此产生的企业资金链断裂、企业破产等产生的经济社会风险和金融风险等隐患,大量人员失业,社会不良资产率上升等带来的社会风险、金融风险上升,其后果在短时间内是体现不出来的,风险发生后将由政府财政来兜底,这部分财政支出也属于隐性支出。

政府"去产能"的机会成本包括但不仅限于在去产能过程中,落后产业的淘汰及产业的优化升级,企业投入增加,利润减少,导致政府税收减少。

去产能过程中各级政府承担的主要成本如图11所示。

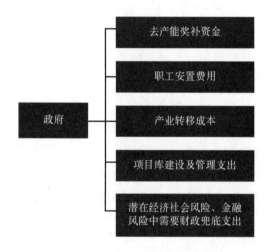

图11 政府"去产能"成本

3. 企业"去产能"成本

在"去产能"过程中,作为市场微观主体的企业面临不同的境遇。有的企业需要整体关停,有的企业需要部分关停。对企业来说,显性成本指在"去产能"过程中企业发生的直接的、当下的支出。市场中部分企业不符合生产工艺、技术、环保、质量等标准,生产水平落后,造成大量的污染,政府对其采取关停措施;另有企业规模较大,其内部已形成生产闭环循环系统,但部分环节仍属于落后产能范围,企业虽然继续存在,但仍将面临部分落后产能的压缩与淘汰。在关停或者企业内部部分关停过程中,都将带来的职工安置、债务处理、资产减值损失等问题;部分关停的企业通过企业间兼并重组、搬迁等举措发挥集聚效应,加速企业转型升级,优化产业结构,减少生产对环境的负面影响。企业在转型升级中,通过增加科研技术投入进行产业升级;为降低工业废弃物对环境的负面影响,企业增加投入,引进清洁设备,对废弃物进行处理后排放。

隐性成本指"去产能"过程中非直接和非当下的支出。例如,"去产能"企业被视同为"落后产能"企业,信用下滑导致借贷成本上升;去产能导致生产过程中原有的循环经济循环链条被打破,部分原材料等不得不从外省采购,导致原材料、运输成本、人工成本等上升。

企业的机会成本,指企业为了响应国家"去产能"政策而丧失的利益选择。例如在市场回暖的情况下,产能压缩后企业丧失市场机会、盈利机会等导致企业收入减少。

去产能过程中企业可能承担的各类成本如图12所示。

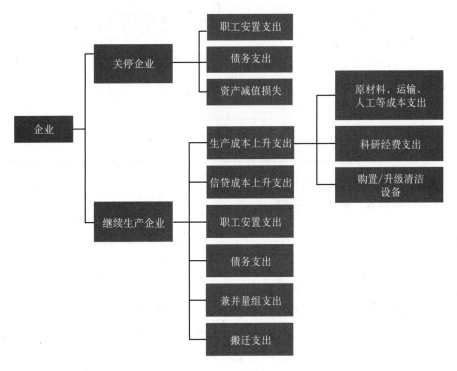

图12　企业"去产能"成本

为了更全面地预估、测算去产能的成本,需要结合现实"去产能"过程中政府、企业采取的行动与措施,合理地进行成本划分。本报告根据各利益主体及其活动特点,采用显性成本、隐性成本、机会成本的概念与分类,以去产能中的利益主体作为横轴,以成本分类作为纵轴,以矩阵形式呈现"去产能"成本(见表12)。

表12 "去产能"成本矩阵

	显性成本	隐性成本	机会成本
政府	1. 以奖补资金(梯级奖励资金)、专项资金(职工安置费用)形式体现的公共支出; 2. 为建立项目信息库和公开制度而增加的管理支出; 3. 政府承担的产业转移成本。	1. 税收优惠 2. 后续财政兜底的社保支出 3. 地方债务风险	由于淘汰落后产能导致的一定时期内财政收入的减少
企业	1. 职工安置成本 2. 兼并重组支出 3. 搬迁支出 4. 科技研发支出 5. 运输成本 6. 原材料成本	1. 企业信用变化导致借贷成本的增加 2. 产业布局链条变化导致的成本增加 3. 企业准入成本	在市场回暖的情况下,产能压缩后丧失市场机会、盈利机会等导致企业收入减少
金融系统(如银行)	对企业的尽职调查成本	潜在的系统性金融风险	

(二)政府去产能的成本分析

根据以上理论分析框架,先结合河北省"去产能"的情况,对政府"去产能"主要成本进行分析。

1. 以奖补资金、专项资金形式体现的公共支出

为推动去产能,中央政府、省政府、市级政府通过多种形式增加财政投入。中央政府通过财政资金奖励的方法支持地方政府去产能工作。2011年4月20日财政部、工业和信息化部、国家能源局印发《淘汰落后产能中央财政奖励资金管理办法》,对"十二五"期间(2011—2015年)各地方淘汰落后产能工作给予奖励,涉及领域包括电力、钢铁、焦炭、酒精、印染等十余项行业。2016年6月14日,财政部印发《工业企业结构调整专项奖补资金管理办法》。为鼓

励地方和中央企业尽早退出煤炭、钢铁过剩产能，设立专项奖补资金1000亿元，实行梯级奖补。专项奖补资金的奖补标准按预算总规模与化解过剩产能总目标计算确定，钢铁、煤炭行业专项奖补资金按照两行业化解产能任务量、需安置职工人数、困难程度等因素确定，并对超额完成任务的省、中央企业实行梯级奖补，专项奖补资金主要用于职工分流安置工作。

为减轻去产能企业的资金压力，河北省及时将国家给予的奖补资金拨付到有关设区市和省直管县（市），2014年还出台《河北省化解钢铁过剩产能奖补办法》，从大气污染防治专项资金中安排8亿元（第一批安排6亿元）、工业转型升级化解钢铁过剩产能专项资金2亿元用于化解钢铁过剩产能，鼓励过剩产能加快退出[①]。按该规定，去除450立方米及以上高炉、40吨以上转炉、30吨以上电炉的，除国家给予的奖补资金外，河北省再按每压减1万吨炼铁产能奖补25万元、每压减1万吨炼钢产能奖补30万元的标准给予奖补。截至2016年10月，已预拨各地省级化解钢铁过剩产能专项资金2亿元。

2016年11月，河北省印发了《河北省化解煤炭过剩产能奖补办法》，按退出民营建设煤矿生产能力进行补助，补助标准按以下两个因素计算：一是每万吨退出产能补助30万元；二是按退出煤矿产能进行梯次奖励，其中，年生产能力6万（含）吨以下煤矿奖励200万元；6万（不含）—9万（含）吨奖励250万元；9万（不含）吨以上奖励350万元。

2016年11月，河北省发布《关于多渠道筹集化解钢铁过剩产能资金指导意见》，按照政府引导、市场运作、企业自愿的原则，拟通过"争取国家支持一部分、各级财政拿出一部分、企业互助筹集一部分"，建立钢铁产能退出补偿机制，以存量产能出资补偿退出产能，促进钢铁产能向有核心竞争力的企业合理流动。目前，化解任务较重的部分市县已经设立化解钢铁过剩产能奖补资金，如唐山2017年安排财政资金1亿元、邯郸市本级2016年安排1.3亿元支持钢铁企业化解过剩产能。

政府从"去产能"的财政专项资金中划出固定比例予以定向安排；对承接再就业员工的企业，中央专项基金与地方过剩行业减产转型基金按照一定比例给予相应的资金奖励与政策倾斜；对下岗分流人员，免费提供再就业职业技能培训；配合国有企业改革，将从债务重组中退出的国有资本充实到社保资金账户中来，稳定和适度增加失业人员的社保资金供给。

---

① http://hebei.ifeng.com/news/chengshi/ts/detail_2014_11/05/3110479_0.shtml。

2013至2016年度中央及河北省政府去产能资金来源投入情况见表13。如表13所示，近年来中央政府去产能资金增长较快，2016年中央政府去产能投入规模已达2013年中央政府投入的15倍。中央政府是去产能公共资金投入的重要来源。

表13　　　　2013—2016年河北省去产能相关奖补资金来源

	中央政府（万元）	占中央本级一般公共决算总支出的百分比（%）	河北省政府（万元）	占河北省本级一般公共预算总支出的百分比（%）
2013	16656	0.008%	10000	0.12%
2014	119622	0.053%	106665	1.31%
2015	113434	0.044%	29776	0.35%
2016	246227	0.090%	104658	0.30%

资料来源：河北省财政厅。

专栏：河北省唐山市"去产能"财政投入状况

唐山市位于河北省的东部，2016年地区生产总值达到6306.2亿元，年均增长6.4%。一般公共预算收入355.1亿元。唐山市的钢铁产量占河北省钢铁总产量的2/3，去产能任务重，政府支持去产能实施中，除了省级以上政府的资金投入，市级政府也加大财政资金整合力度。

唐山市"去产能"坚持淘汰落后产能、产业升级双管齐下，统筹整合唐山市级以上资金10.4亿元，重点支持钢铁、水泥、焦化等行业，减排治理项目；统筹市以上资金11亿元，支持唐山贝氏体钢铁公司等企业化解钢铁过剩产能；投入资金1.5亿元支持中心区燃煤锅炉拆除和集中供热并网工程；统筹市以上资金4.6亿元，全市淘汰黄标车6.2万辆；购置清洁能源、环保公交车515辆；扶持产业发展的各类资金2.25亿元，设立工业、农业、服务业产业投资引导基金和科技风险投资基金，打造强有力的集合平台，助推产业转型升级；投入资金7.5亿元，支持工业企业转型升级，推动钢铁装备制造等产业实施"有中生新"。

按照唐山市大气污染防治工作任务，唐山市统筹安排使用中央、省级和天津市对口支持的大气污染防治资金，唐山市获得大气污染防治专项资金12.5亿元：其中2015年6亿元，2016年6.5亿元，主要用于推进散煤治理，化解钢铁过剩产能，新能源汽车推广，大气污染防治能力建设等大气污染防治工作任务。

加大市级资金投入，2016年唐山市财政安排大气污染防治资金11.4亿元，其中2015年4.8亿元，2016年6.6亿元，用于化解钢铁过剩产能企业补助和对化解过剩产能、淘汰落后产能、大气污染，重点治理等关停企业的援企维稳补助。为促进"去产能"，唐山市出台钢铁去产能资金奖补办法，规定5万元/万吨的奖补标准。

资料来源：唐山市财政局提供。

2. 稳岗补贴资金支出

根据人力资源社会保障部、国家发展改革委等七部门《关于在化解钢铁煤炭行业过剩产能实现脱困发展过程中做好职工安置工作的意见》（人社部发〔2016〕32号）精神，河北省人民政府办公厅出台《关于做好化解钢铁煤炭等行业过剩产能职工安置工作的实施意见》，指出把职工安置作为重中之重，坚持企业主体、地方组织、突出重点、依法依规的原则，因地制宜，分类施策，充分运用市场机制和帮扶措施统筹做好职工安置工作。通过扩大援企稳岗政策、实行内部退养、实行等待退休、促进转岗就业创业、鼓励优势企业兼并重组、运用公益性岗位等服务措施托底帮扶、妥善处理劳动关系、大力开展职业培训、给予特定政策补助的方式进行职工安置。

为支持做好去产能过程中的职工安置工作，河北省政府对于依法参加失业保险并足额缴纳失业保险费，采取有效措施不裁员、少裁员，稳定就业岗位的企业，由失业保险基金给予稳定岗位补贴（以下简称"稳岗补贴"）。各地对符合上述范围和实施条件的企业，在实施产业结构调整和大气污染治理期间，每年可一次性给予稳岗补贴，稳岗补贴所需资金从失业保险基金中列支，稳岗补贴主要用于职工生活补助、缴纳社会保险、转岗培训、技能提升培训等相关支出。稳岗补贴标准为：对于上年度采取有效措施稳定就业岗位且无裁员的企业，可按照该企业及其职工上年度实际缴纳失业保险费总额的50%给予稳岗补贴；对于上年度采取有效措施稳定就业岗位，裁员率低于统筹地区登记失业率满1个百分点的企业，可按照该企业及其职工上年度实际缴纳失业保险费总额的40%给予稳岗补贴；对于上年度采取有效措施稳定就业岗位，裁员率低于统筹地区登记失业率不足1个百分点的企业，可按照该企业及其职工上年度实际缴纳失业保险费总额的30%给予稳岗补贴。

2016年，唐山市拨付稳岗补贴资金10.09亿元，其中向承担去产能，污染治理任务，并符合条件的企业拨付7.71亿元，占比76.4%。2016年以来，当地没有因去产能出现规模性的工人下岗。

3. 产业转移成本

去产能实施过程中，已有产业结构的调整和转移，承接地的产业对接和新产业的培育等，都面临巨大的产业转移成本。在产业转移过程中，需要对相关利益主体进行赔偿、产业转移安置、转移后对原有场地的改造、整理，这些都会带来成本的增加。此外，产业转移后，可能带来承接地产业不能持续发展的成本。可能由于新到地区产业结构差异大，产业的相互依赖性和上下游关联性不够，难以形成产业互动，无法发挥集聚效应、规模效应等，进而导致成本的上升。此外，还会发生公共服务重建成本，新到工业区要根据工人的需要重新规划交通路线、重新建设水、电等基础设施，相应地产生成本。

4. 去产能导致财政收入的减少

短期内，去产能会影响企业的产值和利润，进而影响当地的财政收入。特别是企业关停对当地经济发展影响严重，导致工业产值下降，税收减少。据测算，截止到2017年，河北省因化解产能财政减收74.33亿元，其中钢铁制造业减收63.6亿元，水泥制造业减收9.69亿元，玻璃制造业减收1.04亿元①。2016年，河北省唐山市丰南区地区生产总值617.6亿元，约占唐山市总产值9.8%。2016年丰南区一般公共预算收入为31.5亿元，约占唐山市一般公共预算收入8.9%②。丰南区丰南镇的贝钢公司在去产能中，淘汰炼铁产能105万吨、炼钢产能491万吨，工业产值减少27.2亿元，税收减少1亿元③。

5. 金融风险及社会风险上升导致的财政支出增加

受经济形势下行、市场供过于求、国家"去产能"政策等因素的影响，过剩产能行业的经营举步维艰，部分企业出现严重亏损，银行业坏账率出现上升。在化解债务中，银行是重要主体。若风险处理不当，则会引发系统性金融风险。

多数产能过剩行业是劳动密集型行业，从业人员众多。若淘汰落后产能，关闭僵尸企业，将会导致大量人员失业，社会将面临大量职工转岗和安置的问题，处置不当容易导致企业内部和社会不稳定。例如唐山国丰钢铁公司积极响应国家政策关停工厂北区生产线，4063名职工下岗，企业为保障职工经济补偿金的及时

---

① 数据来源于河北省财政厅。
② 数据来源于《2016年唐山统计快报》。
③ 调研中丰南区财政局提供。

足额发放，多方面筹集资金 3 亿元安置下岗职工。下岗职工的转岗、安置补偿处置不当，不仅影响职工生活，也会增加当地社会不稳定的隐患，潜在的也会极大地增加政府成本。潜在的金融风险及社会风险的防范化解，也可能导致财政支出增加。

（三）企业去产能成本分析

1. 职工安置费用的增加

职工安置伴随去产能的推进，涉及行业的不断增加，任务复杂严峻。在人工成本大幅提升和职工安置资金缺口大的双重压力下，职工内部退养放假轮休等隐性失业风险加剧。安置员工过程中，严格依据《劳动法》《劳动合同法》解除劳动关系支付经济补偿。据了解，2016 年河北当年去产能共安置职工 57785 人，其中转岗安置 32450 人，内部退养 4755 人，终止劳动合同 18295 人，自然减员等 2285 人。据测算，企业安置一名员工需 9 万—20 万元，职工安置可给企业带来巨大的资金压力。

2. 生产运营成本增加

去产能增加了下游企业的运营成本。2016 年以后，河北省钢铁企业多为钢铁联合生产企业，前端的生产被压减后，后端的企业需要外购，直接增加运营成本。主要体现在：河北的水泥、焦炭随着去产能的扩大，已不能满足本省需求，需要从外省购入，运输成本大幅增加。

去产能增加了企业的技术改造、技术研发的成本。去产能不是单纯地去落后产能，更是对产能结构的优化，对环境的治理。相应地，企业在技术研发、安全生产、环境保护、节能减排质量等方面增加投入，提高了生产成本。例如河北省唐山市丰南区钢铁企业共投资 20 多亿元，完成脱硫除尘等大气提标改造工程 152 项；2016 年投资 2 亿元共完成十九台烧结机深度治理工程，确保烧结机及竖炉脱硫、烟气颗粒物排放浓度达到国家规定的特别排放标准。

3. 企业融资成本增加

去产能增加了企业的融资成本。随着国家去产能政策的推行，银行将去产能行业都列为不支持行业，收紧了对该行业企业的信贷规模，加大了企业从银行渠道筹资的成本和难度。其中具有一定优势的、技术领先的企业可能也"受牵连"得不到贷款。银行对企业断贷、抽贷使企业的经营更加雪上加霜。河北一项调查显示，仅邯郸一地银行断贷、抽贷约 75 亿元。断贷、抽贷后企业的贷款来源更为复杂，融资成本提高。

4. 债务处置难度加大

化解过剩产能，企业经营压力加大，效益大幅下滑，运行风险凸显，甚至停产，带来企业应收账款居高不下，三角债形成，因资金严重紧张而导致企业生产经营发展受到较大影响，欠发工资，迟交社保、医保的问题时有发生。去产能后债务处置的难度加大。以唐山市为例，2017年唐山市将有4家钢铁企业产能全部退出，负债总额达54亿元。其中，银行贷款9.8亿元，其余均为企业借款、社会集资、拖欠货款、拖欠职工工资和社会保险等，债务情况复杂，去产能后续债务处置存在潜在的不稳定因素。

5. 搬迁支出

河北省加快推进重大搬迁改造项目，对钢铁焦化等重点污染行业实施搬迁、退城入园，严格按照产能1∶1.25倍减量置换比例。对列入《河北省钢铁产业结构调整方案》的首钢京唐二期、唐山渤海钢铁搬迁、石钢搬迁、永洋特钢和太行、冀南退城进园6个重大项目，组建专门工作组，有针对性地解决制约问题，力促早建成、早达效。制定出台宣钢整体退出方案，连同唐钢、承钢部分产能一并整合重组、减量搬迁。在这个过程中，企业将发生搬迁转移支出。

6. 兼并重组支出

在去产能后，企业因为某些原因无法继续正常运营，经过初步摸底，河北省钢铁行业"僵尸企业"有11家。考虑到员工等各方面利益，通过兼并、重组、破产等方式实现企业的转型发展。在兼并重组中，企业需支付兼并重组的资本投入。

7. 机会成本

企业的机会成本包括部分去产能采取停产限产等措施，但企业阶段性停产，不仅影响企业收入，且增加了企业不必要的人工费用及企业维护费用等。河北省严控新增产能，在市场回暖的情况下，产能压缩后丧失市场机会、盈利机会等，导致企业收入减少。

（四）金融系统去产能成本分析（以银行为例）

1. 坏账率上升导致的成本

据统计，2016年河北省钢铁企业平均负债率65%，2017年负债率可能还会提升。其中，银行和金融机构贷款占有相当的比重。2017年河北省唐山市将有四家钢铁企业产能全部退出，负债总额达54亿元，其中银行贷款9.8亿元。去产能会引起银行不良贷款上升。在产能过剩行业中，低效、高负债的企业占用了大量的信贷资源，一旦资金链断裂就会形成银行不良资产，挤占银行的盈利空间，减少放贷量形成恶性循环，导致系统性风险上升。

2. 对特定行业企业尽职调查成本的增加

由于担心去产能会导致呆账变成坏账,并引致不良资产的上升从而侵蚀利润,银行普遍将钢铁、水泥、燃煤、平板玻璃等河北省重点化解产能行业列入落后产业行业,对该类行业企业收紧信贷额度,停止发放信贷。另外,银行出于防控风险的考虑,加大了对去产能行业相关企业尽职调查,增加了成本。河北一项调查显示,仅邯郸一地银行收缩信贷约75亿元。

三、去产能的效益分析

去产能直观上看只是过剩产能淘汰,实际上是生产结构的优化升级、降低工业排放、提升资源利用效率、应对气候变化等诸多方面,其范围覆盖诸多空间和领域,其绩效具有复杂性和非市场性特征,只从特定方面考察去产能的效益必然有失偏颇,必须从多方面对去产能效益进行评价。本报告拟从生态可持续、社会可持续、经济可持续三个方面对去产能公共支出的效益进行评价(见表14)。在具体评价指标方面,主要利用定量数据对河北去产能的成效进行研究,并将之与全国数据进行对比,对短期内数据不可获得的领域进行定性分析。

表14　　去产能效益分析框架

效益类别	主要体现
生态可持续	减少资源消耗和排放
	实现生态环境质量改善
	推进生态文明建设
社会可持续	促进居民健康
	促进民生发展
经济可持续	提升经济发展质量
	促进经济转型升级
	优化产业布局

(一) 生态可持续分析

去产能的核心目的在于改变当前经济发展过度依赖资源、劳动要素投入的现象,通过结构优化促进整体经济转型升级,打造生态、低碳、绿色、增质、高效的中国经济升级版。就具体资源利用效率而言，去产能涉及资源消耗减少、排放减少及空气质量改善。就整体生态环境效益而言，去产能是推进生态文明建设的重要组成部分。

1. 减少资源消耗和排放

长期以来，中国处于欠发达地位，资源优势是比较优势。但是随着经济发展，资源大量开发，使得资源快速消耗，特别是产能过剩情况下，这种资源消耗更加得不偿失。一方面，当代人消耗掉了后代的资源，而产能过剩却使得这种消耗并不能给当代人带来实际的利益。另一方面，在环保没有跟上的情况下，生态环境遭到了极大破坏。环境承受达到一定域值时，将以加速度遭到破坏，而恢复以前的环境也就变得无比艰难。去产能，削减的是资源利用粗放的无效产能。在传统的产能削减后，新产能、新发展更注重可持续性，注重资源得到合理利用，环境得以保护。

从河北的情况来看，2016年6月河北省政府修订完善了环保、能耗、水耗、质量、技术、安全等6类严于国家标准或行业平均水平的地方标准，倒逼去产能。从目前情况来看，河北相关资源能耗都有显著下降，工业企业能耗指标和万元工业增加值下降速度快于全国水平，要素使用效率显著提高。

从工业企业能耗来看，河北省近几年的工业能耗下降速度较快，除2016年外，这一下降速度均快于全国均值（见图13）。

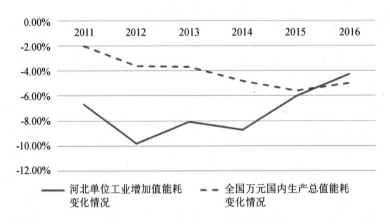

图13　2011—2016年河北与全国能耗变化情况对比

资料来源：历年河北统计公报、历年国家统计局公报。

从万元工业增加值用水量来看，河北一直保持了较高的水资源利用效率，2014年用水量仅为全国水平的29.69%，且水耗下降速度也较快，2013年河北万元工业用水量已排名全国第4，仅居天津、山东、北京之后。2016年河北水利厅还发布了更严格的水资源管理目标，计划至2020年将这一指标压减至12吨（见图14）。

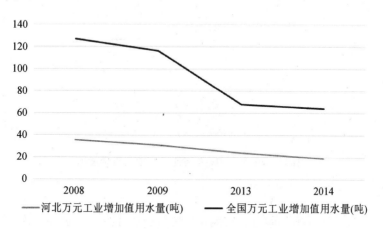

图14 河北近年万元工业增加值用水量变化情况

资料来源：国家发改委网站，国家统计公报，河北日报，河北节约用水规划。

2. 实现生态环境质量改善

良好的生态环境是最公平的公共产品，是最普惠的民生福祉。当前，生态环境是中国的突出短板，补齐生态环境短板是当前经济社会体制改革的重要内容。去产能，即是以产业化治理为抓手，使结构减排、技术减排、工程减排和管理减排相协调。以社会化治理结构为改革方向，形成政府导向、企业主体、公众参与的长效机制，使青山常在、绿水长流，满足城乡居民对宜居环境的期待。

从主要空气指标改善情况来看，河北省2013—2016年年平均PM2.5浓度下降了近40微克每立方米，下降幅度30%以上，年平均下浮幅度在10%以上，其余主要污染物如PM10、臭氧等下降幅度也都较为明显。2016年较2013年空气质量优良天数上升了75天，重污染天数下降了50天。总体来说，空气质量有显著改善（见图15）。

3. 推进生态文明建设

党的十八大以来，习近平总书记站在谋求中华民族长远发展、实现人民福祉的战略高度，按照尊重自然、顺应自然、保护自然的理念，把生态文明建设融入经济建设、政治建设、文化建设、社会建设各方面和全过程。河北省政府自2015年以来，加快推进生态文明建设。去产能是落实生态文明建设的重要方面。从根本上说，去产能就是要改变过去无序利用、浪费资源的开发方式，改变传统

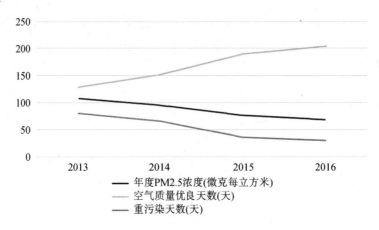

图15 2013—2016年河北相关空气质量指标

资料来源：历年河北省环境状况公报。

要素投入结构过度依赖劳动力、土地和资源等一般性生产要素投入的现象，抑制旧产业、旧业态的供给需求，加快资源从传统三高两低（高污染、高消耗、高危险、低效益、低产出）行业的退出速度，把经济活动过程和结果的绿色化、生态化作为绿色发展的主要内容和途径。通过结构优化促使整体经济转型升级，实现工业化、信息化、城镇化、农业现代化和绿色化同步发展。去产能与生态文明建设的目标一致，从长远看，实施去产能有助于推进生态文明建设。

（二）社会可持续分析

从长远看，去产能促进社会发展可持续。通过挤出落后的过剩产能，推动生产技术升级，改善环境质量，促进居民健康。通过资源合理利用，实现社会可持续发展，有利于实现居民就业的长期稳定，实现居民收入的稳步提升，保障民生发展。

1. 改善生态环境，促进居民健康

良好的生态环境是人类生存和健康的基础，没有全民健康就没有全民小康。环境良好对健康至关重要，只有全方位、全周期确保环境良好，才能有效把控多种健康影响因素交织的复杂局面，并从根本上满足广大人民群众的共同追求。

过去长期以来的经济粗放发展，引致了雾霾等极端天气的出现，对居民生活和健康造成了恶劣影响。去产能通过提高能耗标准、降低排放水平，主动遏制了生态环境的进一步恶化，针对近年来影响居民健康的突出环境问题做出了有力回应。落后产能的淘汰和高效、优质产能的发展，是绿色发展理念的践行。

对近两年河北省呼吸道疾病（肺结核、流行性感冒）的发病情况进行考察，两类疾病的发病规模和总发病数占比，各年度走势都呈现较强的季节性特征，极值差别不大，总体来看年度间差异较小。这一平稳走势，一方面或许说明去产能相关措施有效防止了呼吸道疾病的扩大发展（见图16）。另一方面可能说明去产能对健康的影响是个长期的过程，总体效应短期内并不显著，长期效应有待观察。

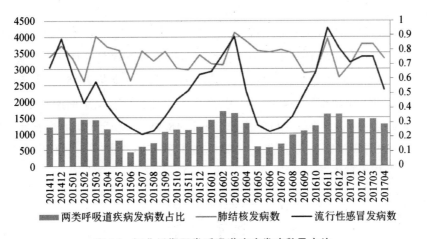

图16 河北近期两类呼吸道疾病发病数及占比

资料来源：河北省卫计委网站。

2. 促进民生改善

从近期看，当前去产能工作推进，河北把维护职工权益、保障职工安置作为工作重点，进一步拓宽职工的分流安置渠道，通过企业内部分流、转岗就业创业、内部退养、公益岗位兜底等安置去产能职工。河北省2016年去产能共安置职工57785人，政府支出有效防止了失业造成的社会冲击。

从长期看，通过淘汰落后的过剩产能，实现资源合理利用，推动建设美丽河北，实现社会可持续发展，有利于实现居民就业的长期稳定，实现居民收入的稳步提升，促进民生改善。

（三）经济可持续分析

去产能是当前中国供给侧结构性改革的重要内容。进入新常态，原有的粗放发展方式存在的内在缺陷逐步暴露、难以为继，必须通过供给侧结构性改革对生产结构进行调整。去产能推动了经济发展质量的提升，推动了经济发展方式的转变，推动了产业布局的优化。

1. 提升经济发展质量

去产能既具有短期调控的政策效应，又具有提升、改善经济发展质量的长期效应。通过直接调整产能结构，能够有效限制生产资料大量投入和工业产品供给过剩价格竞争，对初级资源和工业产品的有序生产能够在短期内发挥效果。去产能通过主动引导产业结构调整，促进生产成本降低、生产效率提高和附加值增加。

就现有情况来看，去产能措施已对河北的产业竞争情况发挥了积极效应，近年规模以上工业相关数据好于全国均值，初步适应了新常态下的经济发展。对比近年规模以上工业相关数据，规模以上工业增加值增速在 2015 年降至最低值 4.4% 后在 2016 年出现了小幅回升，主营业务收入及利润总额增速都由负转正，利润总额增速由 2015 年的 -11% 反弹至增长 18.9%，为近 5 年的最高值（见图 17）。总的来看规模以上工业在经济总量增长不大的背景下，利润出现较大回升，呈现健康发展态势。

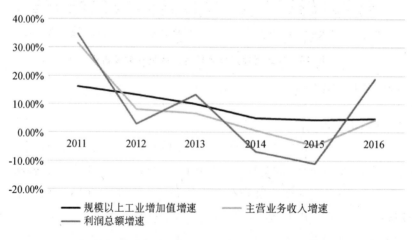

图 17 2011—2016 年河北规模以上工业相关指标

资料来源：河北省统计局网站。

2. 促进经济转型升级

经济转型升级是去产能的重要目的。河北通过去产能，大力推动经济转型升级。

一方面，提高现有存量产能的技术含量。在已执行严于国家标准的地方能耗等生产标准的基础上，鼓励企业加大科技研发投入，推动钢铁工艺装备技术升级、产品质量上档次、节能环保上水平，努力实现钢铁产品细分化、精品化、高端化，争取更多钢铁企业纳入国家级高新技术企业目录；培育壮大建筑用钢结构、金属制品等耗钢产业，积极发展钢材加工配送等非钢产业。力争到"十三五"末，全省钢铁行业主体装备达到国际先进水平，品种钢和特种钢比重明显提高，非钢业务收入占比达30%以上。

另一方面，主动作为引导增量产能提升。以高端化、智能化、链式化、服务化为主攻方向，引导钢铁制造业与"互联网+"融合发展，重点发展交通运输装备、能源装备、工程及专用装备、基础零部件四大产业链，确保装备制造业增加值明年超过钢铁产业。大力发展新技术、新产业、新业态、新模式，实施现代服务业重点行业三年行动计划、高新技术产业倍增计划和科技型中小企业成长计划，加快培育发展新动能，力争到2020年，服务业增加值占生产总值的45%，战略性新兴产业增加值占规模以上工业的20%以上。

从三次产业结构来看，河北近年来三次结构调整效果初显。较2011年，2016年第二产业比值下降16.8个百分点，第三产业占比上升17.8个百分点（见图18）。

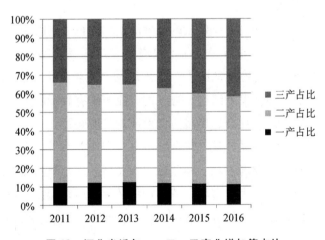

图18　河北省近年一、二、三产业增加值占比

从现有高新技术产业发展情况来看，工业结构转型成效明显。2011—2016年高新技术产业增速一直高于规模以上工业增速，早期两者差距有缩小趋势，但随着规模以上工业增速回落，高新技术产业增速保持在10%以上，两者增速差距逐步拉开，近三年高新技术产业增速始终快于规模以上工业增速8个百分点。

与之相对应，高新技术产业对地区经济的贡献规模在不断增加，2016年已占规模以上工业增加值18.4%，较2015年比例高2.2个百分点，经济转型成效突出（见图19）。

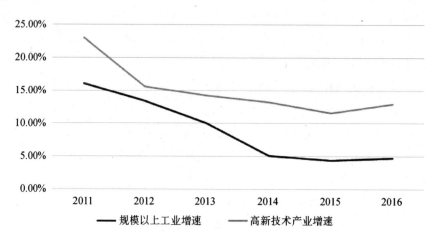

图19　2011—2016年河北高新技术产业增速快于规模以上工业增速

资料来源：河北省统计局网站。

3. 优化产业布局

产业布局的实质是通过产业要素的合理流动和配置达到区域经济结构和产业空间组织结构的优化。产业布局优化不仅有利于产业集聚发展、上下游产业形成，从生态脆弱地区向生态承载能力较强地区的产业转移，还有利于减少污染、节省生态资源，优化经济发展结构。合理的产业布局有利于取得良好的社会、经济、生态效应，是经济、社会、环境协调发展的关键问题。

河北当前面临产业布局分散、整合度不高、产业群水平分工为主、联动力不足等问题，产业布局不合理限制了河北进一步发展。河北坚持将去产能与优布局相结合，通过整合重组、优化布局、境外转移，解决河北钢铁产业集中度低、区域布局不合理问题，以产业集群发展提高产业发展水平和竞争力，为做强钢铁产业奠定基础。

一是整合重组。坚持企业主体和政府引导相结合，支持有条件的企业实施跨区域、跨行业减量重组，提高行业集中度，到"十三五"末形成以河钢和首钢两大集团为龙头、3家地方集团为重点、10家左右特色企业为支撑的"2310"新格局，钢铁企业由109家减少到60家左右。

二是优化布局。按照减量替代的原则，推动环境敏感地区和环京津地区产能有序退出，鼓励引导钢铁产能向沿海临港地区转移或退城进园。"十三五"期间，张家口、保定、廊坊钢铁产能全部退出，秦皇岛、承德原则上按照 50% 的比例退出，其他城市和城市周边的钢厂也要逐步退出或退城进园、向沿海搬迁。

三是推进国际产能合作。鼓励优势企业建立境外生产制造基地，抓好重点合作项目。"十二五"规划期间企业"走出去"和国际产能合作跨越发展，累计完成对外直接投资 70.5 亿美元，是"十一五"的 4.7 倍，河北钢铁集团塞尔维亚 220 万吨钢铁、冀东 120 万吨水泥等一批境外优势产能合作项目顺利实施。

四、主要结论与建议

（一）主要结论

1. 去产能是应对气候变化的重要措施，相关研究亟待加强

旨在加强《联合国气候变化框架公约》实施的《巴黎协定》提出了 2020 年后全球应对气候变化、实现绿色低碳发展的蓝图和愿景，是人类气候治理史上的里程碑。《巴黎协定》的长远目标是确保全球平均气温较工业化前水平升高控制在 2 摄氏度之内，并争取把升温控制在 1.5 摄氏度之内"付出努力"。为实现该协定的长远目标，各方承诺将尽快实现温室气体排放不再继续增加；到 2050 年后的某个时间点，使人为碳排放量降至森林和海洋能够吸收的水平。中国积极履行气候变化治理的责任，加快落实《巴黎协定》将蓝图化为实际行动，明确了二氧化碳排放 2030 年左右达到峰值并努力尽早达峰等一系列行动目标，并将行动目标纳入国家整体发展议程。当前及今后，中国促进经济发展方式向低碳转型是应对气候变化的必然选择。淘汰落后产能是促进经济发展方式转变，提升经济发展质量，应对气候变化的重要措施。去产能的制度、政策、工具、成本效益评估等相关研究亟待加强。

2. "去产能"的高昂成本由政府、企业等相关利益主体分担

为了更全面地评估去产能的成本，需要结合现实"去产能"过程中政府、企业采取的行动与措施，合理地进行成本划分。本报告根据各利益主体及其活动特点，采用显性成本、隐性成本、机会成本的概念与分类，以去产能中的利益主体作为纵轴，以成本分类作为横轴，以矩阵形式呈现"去产能"成本。多维度矩阵分析表明，去产能过程中，政府、企业、银行等各利益相关方付出巨大的成

本代价。在总体成本的分析中，除了关注中央政府、河北省政府以及省内的县市政府安排专项资金支持去产能的显性支出外，还应关注各级政府、企业、银行等相关主体承担的各种隐性支出，尽管本报告对后者很难作出精确的量化分析，但这些支出应得到关注。

3. 去产能在实现经济、社会、生态可持续方面取得明显成效

"去产能"旨在通过工业领域生产方式的变革，实现由粗放经济向循环经济的转变，减少工业废弃物对环境的污染与破坏，降低人类活动对气候的负面影响。对河北来说，"去产能"是生态文明建设的重要组成部分，对改善居民的生活环境、对气候改善有重要意义。去产能产生了良好的社会效益，通过改善环境质量，促进居民健康。通过资源合理利用，实现社会可持续发展，有利于实现居民就业的长期稳定，实现居民收入的稳步提升，保障民生发展。去产能产生明显的经济效益，促进了经济发展质量提升，促进了经济转型升级发展，促进产业结构和产业布局的优化。

4. 从短期看，去产能的成本大于收益；从长期看，去产能的收益大于成本

结合本报告对河北省企业实地调研情况以及本报告对成本收益的分析，从短期看，去产能的成本大于收益，政府、企业、银行等相关利益主体去产能当期的支出压力较大。对河北省去产能行业负债、成本、利润数据的分析表明，去产能导致相关行业企业当前生产成本率提升和利润率下降。但是，从长远看，去产能的收益大于成本。微观层面看，去产能有助于企业降低成本和增加利润。通过提高环保、能耗、质量、安全等生产指标去产能，能够有效剔除落后产能，降低社会整体生产成本。通过调整生产结构，提高生产附加值，最终推动企业利润平稳上升。从宏观层面看，去产能有助于实现经济、社会、生态可持续。

（二）相关建议

1. 去产能需与发展循环经济、优化产业布局统筹考虑

结合对河北省的调研，本报告认为，去产能政策实施中不应一味考虑减排放，而应从大力发展循环经济和废弃物资源化利用、调整和优化产业结构等方面综合考虑去产能的政策设计。

2. 应分类施策，动态优化去产能政策

应根据实际情况，及时调整去产能任务目标，动态优化去产能政策。考虑产业链、产业配套、产业相关，分类施策去产能。如单一的焦化产业可以去产能，但对焦钢一体的企业可另作处理，如考虑安装脱硫设备等。对不同产品区别对

待,例如电解铝中有很多铝产品,钢铁行业中有很多特种钢属于紧缺产品,对于这些产品,政策应引导、支持其提高产品的技术含量和市场竞争力。

3. 中央政府实施的奖补资金政策应予完善

为支持地方化解钢铁、煤炭行业过剩产能,中央财政设立了工业企业结构调整专项奖补资金,用于化解钢铁、煤炭行业过剩产能过程中的职工分流安置。调研中我们了解到,专项奖补资金使用管理还有待完善:调整国家奖补资金分配办法,提高资金使用的灵活性和有效性。奖补资金分配应综合考虑企业化解产能任务,需安置职工人数以及困难程度,可考虑允许企业统筹使用,用于产能补偿、职工安置、弥补资产损失、债务偿还、内部转岗等。在现有规定下,至少应明确奖补资金不仅可用于安置职工,也可由企业用于弥补安置职工的成本,如可列支缴纳职工养老保险,列支补缴社会保险缴费欠费等。在下达奖补资金时,除化解产能关停设备直接涉及的职工外,可把其余相关配套设施所涉及的职工一并考虑,并对提前关停的企业也予以支持。此外,应完善奖补资金的政策绩效评估,以结果为导向,强化政策对企业的激励作用。

4. 气候公共支出成本效益的多维度矩阵分析框架还有待进一步拓展和完善

气候公共支出项目的成本效益分析是气候公共支出评估的重要方面,对于提升公共支出绩效,应对气候变化具有重要意义。本报告以河北省"去产能"为案例,开创性地构建了多维度矩阵分析框架,较为系统地分析了"去产能"的成本与收益。在成本分析方面,区分政府、企业、银行等相关利益主体,从显性成本、隐性成本、机会成本三个方面进行分析;在收益方面,从生态环境效益、社会效益、经济效益等方面进行宏微观分析。需要说明的是,本报告提出的气候公共支出项目成本效益分析的多维度矩阵方法框架,还有待未来通过更多的案例研究予以完善和拓展。

5. 气候公共支出相关基础信息数据的采集和统计工作亟待加强

气候公共支出成本效益分析不仅面临来自技术方法方面的挑战,还面临数据可获得性的挑战。气候公共支出评估研究是个崭新的领域,目前对气候公共支出的统计口径和范围的界定尚待统一,评估方法尚待完善。气候公共支出涉及的部门众多,成本与收益的基础信息数据的生成需要众多政府部门的配合。未来,应进一步加强气候公共支出基础信息数据的采集和统计工作,加快国家气候变化基础信息数据库建设,整合分散在发改、财政、环保、农业、卫生计生、统计等部门的数据和信息资源,实现基础信息的互联互通、动态更新和综合集成。

结　　论

随着世界步入以可持续发展目标为新发展纲领的时代，绿色发展理念在最新《中华人民共和国国民经济和社会发展第十三个五年规划纲要》中得到了高度重视。为实现绿色发展，中国政府提出了生态文明建设这一总体框架，以整合各项政策工具，引导各项资源，集中社会各界力量来共同推动具有可持续性和包容性的均衡的经济增长。

在这一背景下，《中国气候公共支出分析与评估——基于河北省的研究》这一项目旨在进行省级层面的制度结构、现有政策和公共财政支出的分析，从而对现阶段中国采取的各种应对气候变化的行动进行系统的评估，并衡量不同要素之间的互补程度。而其最终目标则是明确下一步政府干预的关键性领域，并为加强中国气候变化治理提供潜在解决方案。

本报告以中国河北省这一存在着严峻环境问题的重工业省份为研究目标。鉴于其在结构化经济转型和实现绿色增长目标等方面面临的挑战，河北省是检验生态文明建设方式的一个贴切的案例。对那些既致力于保持经济增长势头、又致力于探索减缓自身气候恶化和保持环境完整性道路的其他省份或其他发展中国家的次国家政府而言，河北省在这方面可为他们提供宝贵的经验。

1. 气候公共支出与制度评估

由于赋予了绿色增长以战略意义，河北省在气候变化的应对方面已取得显著进展。在制度层面，自 2008 年起，该省设立了应对气候变化工作领导小组这一跨部门协调机构，专门负责应对气候变化相关活动的开展。这一举措确保了各个机构以不同方式同步的参与到有助于缓解气候变化的任务中来。总的来说，不同行政系统的部门都根据其能够履行的不同职能付出了应对气候变化的努力,包括气候缓解和适应、技术开发和能力建设及国际合作等。

在应对气候变化工作领导小组的总体领导下,河北省已经发布了50多份旨在应对气候变化的政策文件,主要聚焦于二氧化碳排放的缓解(如提高能源效率、优化能源结构、增加森林碳汇等)和适应(如开展水资源管理、农林、海洋资源和基础设施、灾害风险管理等领域的能力建设)。这些文件都以指导省内发展和保护环境为主要目的。

为实施这些政策,河北省调动了各种资金来源为其提供支持。比如,该省为有助于应对气候变化的9大领域提供了财政支持,涉及生态恢复、空气污染控制、水资源管理、能源效率和农业发展等方面。值得一提的是,该省已将许多创新性融资方式(如公私合作)运用于应对气候变化问题。这意味着,更多资金来源已被用作财政资金的补充,以扩大气候融资的规模。

报告发现,"十二五"期间,河北省气候变化公共支出一直呈上升态势。2011年,河北省一般公共预算支出的8%左右用于与气候变化高度或中度相关的活动。这一数字每年都在上升,2015年已达到11%左右,相当于390亿元(61亿美元)的财政支出。而且,值得注意的是,河北省的气候公共支出水平高于中央政府的支出水平。比如,河北省预算的10%左右用于与气候变化高度或中度相关的活动上,而同时期中央层面支出为7%。

2. 成本效益分析

报告还就河北的去产能行动进行了成本效益分析。为此,报告建立了一个总体理论分析框架,从多维度对去产能的成本和效益进行了考察。具体而言,报告定义并明确了每一利益相关方在去产能活动中的显性成本、隐性成本和机会成本。

对政府而言,去产能的主要成本包括税收减少、为职工安置提供补贴等。就后一项而言,河北省2016年的支出就达人民币1.05亿元(相当于1600万美元)。对于被安排去产能任务的企业,去产能的主要成本则涉及失业人员的安置费用和产业升级相关费用等。据估计,重新安置一名员工要花费一个公司9万元至20万元不等的费用。2016年,河北省共安置57785名下岗工人,共花费人民币520万元至1160万元(相当于80万美元至180万美元)。

在效益方面,河北省的3个可持续性核心指标已全部显现出积极效果:工业企业的能耗水平持续下降;经济结构正在转型和优化(比如,与2011年相比,

2016年第三产业的GDP占比已经上升了17.8%）；空气质量的提高很可能促进了地方居民健康状况的改善（尽管这一效果短期内难以显现）。

3. 下一步行动

报告试图对中国省级层面的气候制度、公共支出及其成本效益进行评估。除河北省相关实证发现外，报告还有两大理论方面的贡献。

首先，对预算项目的气候相关性进行归类时，报告还深入考察了相关活动的性质，即属于气候减缓活动还是气候适应活动。这种细分既有助于更好地了解气候支出的潜在影响，又有助于对气候支出目标活动组合进行广泛追踪，从而为创设、调整或平衡预算项目以实现预期成果最大化的决策提供了有益的视角。比如，可对同时促进气候缓解和适应的高度相关活动计划更多预算。

其次，报告尝试性地创建了一个用于指导气候支出相关的成本效益分析的全面理论框架。这是设立多重标准，从可持续性三大支柱的视角进行成本效益评估的最早尝试之一，指出了不同发展重点（如经济增长、社会保障和环境保护等）和不同利益相关方利益之间的平衡所要面临的挑战。而且，这个框架还使人们注意到了系统性方法的重要性——这一方法强调的是不同产业之间的相互联系和特定产业内部的纵向发展。比如，计算成本时，将整个产业链上可能受气候活动影响的货物和服务的供需情况纳入考量是必不可少的。

报告同时也为一些问题的深入研究奠定了基础。首先，有关气候公共财政支出的成本效益分析，例如随着数据收集的增加，未来可以对"具体支出的收益比"这一问题进行深入调查。其次，在对创建一体化国家发展融资框架的准备过程中，可对更多国内资金来源进行分析和整合。在中国背景下，绿色金融相关分析的拓展更有其必要性，因为中国运用了更加广泛的资金流（如绿色债券、保险、排放交易机制等）来推动绿色发展，其中包括对气候变化的应对等问题。在这一过程中，私营企业起着至关重要的作用。为实现更好的发展效果，它们的资金贡献需要更好的监测和评估。

附件（一）河北省气候相关公共支出政策

本部分依据公开可获取的资料梳理收集了与应对气候变化相关的财政支出政策。相关财政支出信息分为财政投入政策和创新支出方式两大类，财政投入政策又可分为直接应对气候变化、生态修复、地下水超采治理、节能减排降耗、优化能源结构、能效提升、农业综合开发、农村改造、防汛抗灾9类；创新支出方式又可分为政府和社会资本合作、国际贷款、清洁发展基金、排放权交易4类。

1. 财政投入政策

（1）直接应对气候变化

2013年至2015年，河北省共投入457.3亿元治理大气污染（各年度依次为120亿元、160.3亿元、177亿元），并在资金分配、监管、评价机制方面不断创新，保障了各项治污工程顺利实施。通过整合预算资金，调整预算安排，统筹省级环境保护、大气污染防治、工业企业技术改造、节能减排、战略性新兴产业、科技创新等相关专项资金和国有资本经营预算，优先支持大气污染防治相关支出。

（2）生态修复

2014年，河北省着力抓好地下水超采综合治理试点工程、绿色河北攻坚工程、南水北调中线配套工程、引黄入冀补淀工程、山体修复工程、尾矿库综合治理工程、北戴河及相邻地区近岸海域污染治理工程、白洋淀综合治理工程、湖泊湿地保护工程、农业节水灌溉工程、重污染河流治理工程、双峰寺水库建设工程、水土流失治理工程、高标准农田建设工程、农业清洁生产示范区建设工程、京津保生态过渡带建设、河流水网建设、张承水源涵养生态功能区、张家口坝上地区退化林分改造工程等项目建设，总投资约700亿元。2016年，为保护林地，停止天然林商业性采伐，国有天然林停伐补贴按照国家林业局核定的停伐量，以每立方米1000元标准补助，国有的天然林管护补助标准为每年每亩6元，集体和个人所有的天然林管护补助标准为每年每亩15元。

(3) 地下水超采治理

河北省的地下水超采量和超采面积均占全国的1/3，从2014年起，国家在河北省开展地下水超采综合治理试点。明确水权，制定水价，大力实施水价综合改革；控制总量，强化管理，从严管控地下水开采使用；节约当地水，引调外来水，着力发展现代节水农业和增加替代水源；通过综合治理，压减地下水开采，修复地下水生态。2014年在4市49个县开展地下水超采综合治理，安排投资74.9亿元。2015年试点扩大到5市63个县，安排投资82.6亿元。2016年试点范围扩大到9个设区市、2个省直管县，共115个县（市、区），安排投资87.12亿元。2014年度、2015年度试点已形成农业压减地下水开采能力15.2亿立方米。

《河北省地下水超采综合治理试点方案（2016年度）》明确，政府对地下水超采治理项目给予一定补贴，引导农民和农业经营主体积极参与调整种植结构和采取各项节水措施。对调整种植模式项目，按每亩500元给予补助。对旱作农业项目，按每亩100元给予补助。对非农作物替代农作物项目，按亩均1500元给予补助（连续补5年，第二年开始减半）。对推广冬小麦节水配套技术，按亩均75元给予节水品种物化补助。对喷灌、微灌、水肥一体化等高效节水项目，按亩均1500元（含灌溉计量设施）给予综合补助。对管道输水项目，按亩均870元（含灌溉计量设施）给予综合补助。对保护性耕作项目，按每亩50元补助农机作业费。

(4) 节能减排降耗

2012年出台《河北省节能减排"十二五"规划》，明确了"到2015年，全省万元GDP能耗将比2010年下降18%"等目标。2013年，大气污染防治专项资金列支淘汰落后产能4.9亿元。对钢铁、水泥、玻璃、煤炭、造纸、印染等高耗能、高污染和落后技术、产能过剩企业和项目实施了淘汰关停，全年共关停取缔企业8347家，压减粗钢产能788万吨、炼铁586万吨、水泥1716万吨、平板玻璃1488万标准重量箱。通过淘汰关停，进一步优化了经济结构，减轻了环境治理压力。大力推进"6643工程"，即到2017年压减6000万吨钢铁、6000万吨水泥、4000万吨煤和3000万标准重量箱玻璃。其他相关的工程还包括新老"双三十"、"双千"、重点节能行业关键实用技术创新、压减钢铁过剩产能"周日行动"、压减水泥过剩产能集中行动等。针对"煤改气"、粘土砖瓦窑取缔、"拔烟囱"、黄标车淘汰等专项行动，河北省通过财政预算调控，对超额完成治理任务的地方，省财政厅在资金分配时给予重点资金奖励，激励和调动了地方政府防治污染积极性，确保了省政府年度重点工作目标的按时完成。

（5）优化能源结构

在组织实施集中供热、"煤改气"、"煤改电"、洁净型煤推广、燃煤锅炉置换等能源替代工程的同时，大力推广使用洁净煤、型煤、生物质能、太阳能、地热等清洁能源。

燃煤是大气污染重要来源，河北省 2014 年共投入 8.1 亿元对燃煤锅炉实施提前淘汰和节能环保提升改造，改造后达到二级及以上能效标准的锅炉，按每蒸吨不超过 2 万元进行奖补，单个项目补助资金不超过总投资的 50%。对拆除取缔、置换调整、更新换代等方式实施燃煤锅炉淘汰的，也按每蒸吨不超过 2 万元给予奖补。

2015 年开始实施燃煤锅炉治理，到 2017 年底河北省要完成 11071 台燃煤锅炉淘汰任务。对保留的 23562 台燃煤锅炉，按照 2015 年、2016 年、2017 年各完成 30%、30%、40% 的计划，确保按质按量完成节能环保综合改造提升。

2015 年河北通过开展燃煤机组超低排放升级改造、关停取缔实心黏土砖瓦窑和燃煤锅炉治理等，全省削减煤炭消费 500 万吨。目前已淘汰燃煤小锅炉 12009 台、近 3 万蒸吨，建成区 10 蒸吨及以下燃煤锅炉已全部淘汰。完成燃煤机组超低排放改造 252 台。关停取缔实心黏土砖瓦窑 2780 座、彻底解决 260 万吨燃煤低空直排问题。

2016 年 4 月印发《河北省焦化行业污染整治专项行动方案》，提出至年底通过推出转型，压减煤炭产能 600 万吨，确保现有焦炭生产企业全部达到污染物排放要求。

2016 年 4 月印发《河北省露天矿山污染深度整治专项行动方案》，提出利用 3 年时间，对全省 1881 个露天矿山污染进行深度整治。在按比例退还矿业权价款和恢复治理保证金基础上，对 2016 年 6 月底前主动申请关闭的露天矿山给予适当补偿和奖励，对因违法违规被强制关闭的有证露天矿山，不退还矿业权价款和恢复治理保证金，也不给予补偿。

2016 年 6 月印发《河北省散煤污染整治专项行动方案》，提出利用三年时间，对全省散煤生产、流通、使用等环节进行综合整治。

2016年河北省政府出台《关于加快实施保定廊坊禁煤区电代煤和气代煤的指导意见》,为改善生态环境和提高生活质量,到2017年10月底前,禁煤区完成除电煤、集中供热和原料用煤外燃煤"清零"。为加快推进农村地区"电代煤""气代煤",兼顾群众承受能力,农村实施"电代煤""气代煤"将享受到一系列补贴政策,包括设备购置补贴、用电用气补贴、不再执行阶梯电价和气价等。其中,"电代煤""气代煤"设备购置分别补贴85%、70%,每户最高补贴金额分别不超过7400元、2700元;采暖期居民用电补贴0.2元/千瓦时,每户最高补贴电量1万千瓦时,用气补贴1元/立方米,每户每年最高补贴气量1200立方米。

2015年6月1日起施行《河北省人大常委会关于促进农作物秸秆综合利用和禁止露天焚烧的决定》,在原有河北省秸秆露天禁烧专项资金的基础上,另设秸秆综合利用专项资金,重点支持秸秆机械化粉碎还田、秸秆青贮饲用、秸秆收集储运服务体系建设、生物质炉具推广以及秸秆气化、固化成型等资源化利用。

(6)能效提升

推动工业技改,促进转型升级。2013年投入工业企业技术改造资金10.5亿元,对重点改造和建设项目实施银行贴息和专项支持,全年共推动重点改造和建设1124个,引导工业技改投资7200亿元,促进了战略新兴产业[①]、传统优势企业和现代服务业发展。

2011—2013年,河北共有594个工业企业节能减排增效技术项目纳入国家和省级各类专项资金支持范围,其中包括388项节能减排降耗技术改造项目,39个河北省重点建设工业企业能源管理中心,43项清洁生产技术改造示范项目,115个资源综合利用技术改造项目,58项节能环保产品技术改造项目,获得国家和省级资金支持合计23.3亿元,项目的实施有效缓解了工业主要污染物排放量上升趋势、改善了环境质量。

2013年,河北省财政厅共筹措专项资金5600万元,重点支持了28个企业创新能力建设项目。为加大对战略性新兴产业的支持力度,河北省从2012年开始设立战略性新兴产业专项资金,省财政每年安排10亿元,专项用于支持战略性新兴产业关键技术研发、高新技术成果产业化、创新能力建设、重点应用示范、高成长性企业培育、重大龙头项目引进、产业创新发展以及区域集聚发展等

---

① 战略性新兴产业是2010年中央提出的对于推进产业结构升级和经济发展方式转变、提升中国自主发展能力和国际竞争力、促进经济社会可持续发展、具有重要意义的七个产业,具体包括节能环保、新一代信息技术、生物、高端装备制造、新能源、新材料和新能源汽车七类。

工作。2013年，省级财政预算安排了央企进冀及战略性新兴产业、现代物流业发展发展专项资金共计16亿元。2014年省财政安排重点产业发展专项资金43.8亿元，通过财政扶持政策支持冀中南经济区大力发展先进制造业和县域特色产业，努力形成分工明确、优势互补、各具特色、协调发展的产业格局。

通过河北省战略性新兴产业专项资金、现代物流业发展专项资金、央企进冀专项资金，着力发展市场潜力大、产业基础好、带动作用强的行业，加快形成支柱产业，突出科技创新和新兴产业发展，加快转变经济发展方式。

2014年，河北省用于大气污染防治和节能减排的研发资金达到5000多万元，还依据"135"工程优化配置省内外科技资源，逐年增加科技资金投入，按照先急后缓的原则每年布局一批重大科技项目，不断突破大气污染防治技术瓶颈，为压减燃煤、控车减油、治污减排、清洁降尘提供全面科技支撑。

（7）农业综合开发

近几年，河北省财政每年对扶持粮食生产的投入增幅大都在30%左右。2013年，河北省发放68.3亿元种粮补贴，投入财政资金17.38亿元支持农田水利基础设施建设，解决1909万亩耕地浇水问题；投入农业救灾资金4.47亿元，抢险救灾、水毁水利工程修复、农业生产救灾和设施恢复重建；发放小麦、玉米、水稻、棉花等农作物良种补贴资金12.6亿元，兑付农机购置补贴10亿元；下达资金2.65亿元，对2011年粮食总产增幅3.4%以上的77个粮食生产先进县市区实施奖励，专项用于支持粮食生产。安排农业综合开发土地治理项目省级补助资金4.43亿元，用于中央农业综合开发土地治理项目配套。安排中央小农水重点县项目省级配套4.16亿元，用于中央安排的小农水重点县项目建设；省级财政安排2.5亿元，在全省开展1000万亩农田深松作业；安排现代农业项目县省级补助资金1.63亿元，用于与中央安排的现代农业生产发展项目配套；筹集农业高产创建中央补助资金1.1亿元，支持小麦、玉米、大豆、马铃薯、杂粮等高产创建万亩示范片等。

2014年，财政下达市县惠农补贴资金68.3亿元（其中粮食直补7.8亿元、农资综合补贴60.5亿元），补贴面积7890万亩。

2015年，中央安排的农业综合开发资金达到20.24亿元，同时，通过贷款贴息、财政补贴等多种方式，积极吸引社会资金。全省农发财政资金总投入达到27.48亿元。省级财政安排专项资金5000万元，改造提升建设省级现代蔬菜产业园100个。自2016年1月1日起对产粮大县稻谷、小麦、玉米三大粮食作物农业保险保费财政补贴比例进行调整，提高中央财政和省级财政补贴比例，市级财政补贴比例不变，县级财政补贴比例降至零。自2016年7月1日起，河北省对特色农业产业、特色农业产品开展保险保费财政奖补试点。

(8) 农村改造

为促进城乡一体化和公共服务均等化，2014年河北省提出建设3000个美丽乡村，全面完成改造提升行动的15件实事。河北省下达省级补助资金1.43亿元，用于重点村规划编制、垃圾清运和试点村民居节能改造，对确定的10个试点村共4417户，安排试点村民居节能改造专项奖补资金，奖补标准为每户5000元，合计2208.5万元。

2014年，河北被确定为中央补助地方"粮安工程"危仓老库维修专项资金重点支持省份，河北省计划投入12亿元（其中利用中央补助3亿元），对11个设区市、151个县（市、区）的国有粮食企业仓房进行维修改造。

2015年以来，河北省加大了对贫困地区农村危房改造支持力度。农村危房改造资金向7366个贫困村所在县重点倾斜，同时兼顾其他县（市、区）农村面貌改造提升重点村、地震高烈度设防地区等区域。截至2015年9月，河北省已下达农村危房改造资金16亿元，其中中央资金10.45亿元，省级配套资金5.58亿元。已支持完成农村危房改造任务12.3万户，其中建筑节能示范户3万户。

(9) 防汛抗灾

为做好防汛抗洪抢险救灾工作，减轻"7.19"洪涝灾害影响，2016年河北省共下拨中央和省级救灾资金25.87亿元，为灾区群众救助和灾后恢复重建工作提供了有力的资金保障。

2. 创新支出方式

(1) 政府和社会资本合作

2015年4月30日河北省发布了首批交通能源市政等领域鼓励民间投资项目清单，涉及高速公路、一级公路、铁路、清洁能源、热电联产、水力发电、军民品生产、医疗设施，以及城市供水、供热和污水处理等方面，共38项，总投资2106.1亿元，鼓励民间资本以合资、独资、参股、特许经营等方式参与项目建设及运营。2016年5月，河北省物价局、省住房和城乡建设厅出台《关于城市地下综合管廊实行有偿使用制度的实施意见》，要求各地按照既有利于吸引社会资本参与管廊建设和运营管理，又有利于调动管线单位入廊积极性的要求，建立健全城市地下综合管廊有偿使用制度。意见自2016年6月1日起实施，有效期5年。至2016年底，河北省项目库储备项目529个，总投资1万多亿元，其中落地项目42个，投资1636亿元。

（2）国际贷款

2015年12月10日，亚洲开发银行执董会批准了河北省大气污染防治项目，提供3亿美元政策性贷款，用于开展"京津冀空气质量改善——河北政策改革项目"。这是亚行首次向中国发放政策性贷款。德国复兴信贷银行还将提供1.5亿欧元联合融资与本笔贷款捆绑使用，共同推动京津冀大气污染防治。2016年6月6日，世界银行执董会正式批准了河北省结果导向型贷款大气污染防治项目，贷款额5亿美元，贷款期限19年，是国内首个世行结果导向型贷款项目。贷款拨付后，将与亚洲开发银行3亿美元政策贷款及德国复兴信贷银行1.5亿欧元促进贷款一同设立股权投资基金，充分发挥资金的引导放大作用，撬动更多社会资本投向河北省大气污染防治项目及相关产业。

（3）清洁发展基金

国务院2007年批准成立中国清洁发展机制基金，河北省于2011年起就开展了清洁基金有偿使用工作，是最早使用这项贷款的省份之一。截至2014年年底，河北已利用清洁基金优惠贷款5.95亿元。

（4）排放权交易

2013年，河北省财政厅联合省环保厅大力推广排污权交易，积极探索挥发性有机化合物排污收费和重污染企业环境保险政策，对重点行业清洁生产示范给予引导性资金支持。在这项政策的支持下，企业不仅获得了财政资金的支持，还能够得到金融机构的贷款支持。2013年，全省13家有偿获得排污权的企业，得到光大银行石家庄分行提供的排污权质押贷款3444.8万元。

2014年12月，京冀跨区域碳排放权交易市场率先启动建设，交易市场将建立跨区域统一的核算方法、核查标准、交易平台等，意在为国家推行碳排放权交易制度、建设统一碳市场探索经验。跨区域碳排放交易市场实行二氧化碳排放总量控制下的配额交易机制，交易产品包括碳排放配额和经审定的碳减排量（核证自愿减排量、节能项目和林业碳汇项目产生的碳减排量）。两市的市场交易主体可自由买卖排放配额和经审定的碳减排量并可用于履约。

附件（二）河北省气候相关政府部门

河北省气候相关政府部门及其职责

序号	相关政府部门	与应对气候变化相关的主要职责	相关处室
1	河北省发展和改革委员会	——发展循环经济的规划和政策措施，并协调实施；参与编制环境保护规划，协调制订全省应对气候变化工作；指导散装水泥推广和节能监察工作。 ——组织编制农业和农村发展、生态环境建设。 ——统筹能源、交通运输发展规划与资源、环境协调发展。 ——综合分析全省经济社会发展与国民经济和社会发展规划、计划的衔接平衡。 ——能源发展改革中的重大问题；参与研究能源消费总量控制目标建议，提出能源发展战略建议；负责煤炭开发、煤层气、煤炭加工转化为清洁能源产品的发展规划、计划和政策并组织实施。 ——拟订火电、核电和电网（不含农村电网）发展规划，指导协调新能源、可再生能源综合利用；监督能源消费总量控制目标实施。 ——负责指导能源行业节能和资源综合利用；负责石油、天然气、炼油、煤制天然气、煤制燃料和生物质液体燃料的行业管理。 ——组织拟定能源市场发展规划和能源市场设置方案；监管能源市场运行；处理能源市场纠纷；研究提出调整能源价格建议。	应对气候变化处、农村经济处、基础产业处、能源局综合处、能源局煤炭处、能源局新能源处、能源局能源市场监管处

84

附件（二）河北省气候相关政府部门

续表

序号	相关政府部门	与应对气候变化相关的主要职责	相关处室
2	河北省财政厅	——综合管理国家财政收支、财税政策，实施财政监督，参与国民经济进行宏观调控。 ——拟订国际金融组织、外国政府和清洁发展委托贷（赠）款管理制度，负责国际金融组织、外国政府和清洁发展委托贷等贷（赠）款的申请、转贷、偿还、资金使用、统计等工作；承办涉外经济合作与交流事务。	税政处、社保处、资环处、经建处、预算处、农业处、综合处、PPP办、采购办
3	河北省环境保护厅	——负责建立健全环境保护基本制度。拟订并组织实施全省环境保护政策、规划，起草地方性法规和规章草案。组织拟订并监督实施环境保护重点区域、流域污染防治规划和饮用水水源地环境保护规划，按省政府要求会同有关部门拟订重点流域污染防治规划，参与制定全省主体功能区划。 ——负责重大环境问题的统筹协调和监督管理。预案下重大突发环境事件的应急、协调和监督海洋环境保护工作。指导、协调解决有关跨区域环境污染纠纷，统筹协调全省重点流域、区域、海域污染防治工作。 ——承担落实全省环境保护领域固定资产投资规模和排放总量控制目标的责任。组织制定主要污染物排放总量控制和排污许可证制度并监督实施。 ——负责提出环境保护领域固定资产投资规模和方向、省级财政性资金安排的意见，参与指导和推动全省循环经济和环保产业发展，参与应对气候变化工作。 ——承担从源头上预防、控制环境污染和环境破坏的责任。受省政府委托对重大经济开发计划进行环境影响评价，对涉及环境保护的地方性法规草案提出国家和省发展规划以及重大经济开发计划进行环境影响评价，对涉及环境保护的地方性法规草案提出意见，按国家规定审批省重大开发建设区域、项目环境影响评价文件。 ——负责污染防治的监督管理。制定水体、大气、土壤、噪声、光、恶臭、固体废物、化学品、机动车等的污染防治管理制度并组织实施，会同有关部门监督管理饮用水源地保护，组织指导城镇和农村的环境综合整治工作。	机动车污染管理处、大气污染防治处、水污染防治处、农村环境保护处、环境监测与应急处、辐射安全管理处（核安全管理处）、污染物排放总量控制处、政策法规处

续表

序号	相关政府部门	与应对气候变化相关的主要职责	相关处室
3	河北省环境保护厅	——指导、协调、监督生态保护工作。拟订生态保护规划，协调和监督生态破坏恢复工作。指导、协调、监督对生态环境有影响的自然资源开发利用活动，监督对各种类型的自然保护区、风景名胜区、森林公园的环境保护工作，协调和监督野生动植物保护、湿地环境保护、荒漠化防治工作。协调指导农村生态环境保护，监督生物技术环境安全，牵头生物物种（含遗传资源）工作，组织协调生物多样性保护。 ——负责环境监测和信息发布。制定预警、预测预报制度，组织实施省级环境质量监测、污染源监督监测，建立和实行环境质量公告制度，统一发布全省环境综合性报告和重大环境信息。 ——开展环境保护科技工作，组织环境保护科学研究和技术工程示范，推动环境技术管理体系建设。 ——开展环境保护对外合作交流，研究提出国际、省际环境保护合作中有关问题的建议，组织协调环境保护国际条约的履约工作，参加处理涉外环境保护事务。 ——组织、指导和协调环境保护宣传教育工作，制定并组织实施环境保护宣传教育纲要，开展生态文明建设和环境友好型社会建设的有关宣传工作，推动公众和社会组织参与环境保护。	
4	河北省工业和信息化厅	该处主要负责拟订并组织实施全省工业的能源节约和资源综合利用、清洁生产促进政策、参与拟订能源节约和资源综合利用、组织和指导工业行业节能管理；提出全省工业行业需要淘汰的落后工艺、设备（产品）制定、企业节能管理；组织协调相关政府审批和核准投资项目的能耗、水耗审核意见、新设备、新材料的推广应用；研究制定并组织实施工业"三废"资源的综合利用政策及项目管理。	节能与综合利用处
5	河北省林业厅	——组织、协调、指导和监督全省造林绿化工作。制定全省造林绿化指导性计划，指导植树造林、封山育林和以植树种草等生物措施防治水土流失工作，指导监督全民义务植树和造林绿化的相关工作。 ——承担全省森林资源保护发展监督管理的责任。	造林绿化管理处、森林资源管理处、政策法规处、森林资…

续表

序号	相关政府部门	与应对气候变化相关的主要职责	相关处室
5	河北省林业厅	——组织、协调，指导和监督全省湿地保护工作。拟订全省湿地保护规划，实施建立湿地保护区、湿地公园等保护管理工作，湿地保护的有关地方标准和规定，组织、监督和指导全省荒漠化防治工作。组织拟订全省防沙治沙规划，指导沙化土地的合理利用，组织指导全省荒漠化、石漠化防治及沙化土地封禁保护区建设规划，监督沙化土地的合理利用，组织指导对沙尘暴灾害预测预报和应急处置。——负责全省林业系统自然保护区的监督管理。在国家和全省自然保护区规划、规划原则的指导下，依法指导森林、湿地、荒漠化和陆生野生动植物类型自然保护区的建设管理和监督，监督管理林业生物种质资源，植物新品种和生物多样性保护。——指导监督全省产业对森林、湿地、荒漠利用生野生动植物资源的开发利用。——参与拟订林业及其生态建设的经济调节政策，组织指导林业及其生态建设的生态补偿制度的建立和实施，编制部门预算并组织实施，监管国有林业资产和森林资源资产，指导监督全省林业资金的管理和使用。	湿地资源管理处、政策法规处、规划与资金管理处
6	河北省农业厅	——负责农业资源保护工作。指导农业用地、渔业水域、草原、宜农滩涂、宜农湿地以及其生态农业生物质产业发展和农业农村节能减排，承担农业面源污染治理有关工作；划定农产品禁止生产区域，农业生态农业、循环农业等发展；负责保护渔业水域生态环境。——制定并实施生态农业与农村可再生能源建设规划、指导农业可再生能源开发与利用。	美丽乡村建设处、现代农业园区处（山区综合开发办公室）、农业资源环境处
7	河北省气象局	在本行政区域内组织对重大灾害性天气跨地区、跨部门的联合监测、预报工作，及时提出气象灾害防御措施，并对重大气象灾害作出评估，为本级人民政府组织防御气象灾害提供决策依据；管理本行政区域内公众天气预报、灾害性天气预报以及有关气象预警、火险气象等级预报及气象信息的发布，城市环境气象预报，保护气候资源应用推广气象灾专业成果和推广应用；组织气候变化影响评估，组织开展气候变化工作，技术开发和决策咨询服务。——负责本级气候资源开发利用和有关部门项目进行气候可行性论证；参与省政府应对气候变化工作。	

87

续表

序号	相关政府部门	与应对气候变化相关的主要职责	相关处室
8	河北省住房与城乡建设厅	——拟订城市建设和市政公用事业的中长期规划、改革措施、规章制度和技术标准，热力、市政设施、园林、市容环境治理、城建监察；指导城市供水、节水、燃气、配套建设、指导城市绿化工作；指导城市垃圾处理和配套设施建设；指导城市规划区的绿化工作；负责省级以上风景名胜区的审查报批和监督管理。 ——拟订建设机械材料设备行业的政策和发展规划并监督实施；指导房屋墙体材料革新工作；指导和规范建设机械材料设备行业的有关规范；组织定期发布建设机械材料设备淘汰、限制使用和推广应用产品目录；指导和管理建筑节能和建筑节能材料产品使用工作。	城市建设处、建设材料装备处
9	河北省水利厅	——按照国家资源与环境保护有关法律法规和标准，拟定水资源保护规划；组织水功能区划分和向不同功能区水域排污的控制；监测江河湖库水量、水质，审定水域纳污能力，提出限制排污总量的意见。 ——拟定全省水利经济调节措施；对水利资金的使用进行宏观调节；贯彻执行国家有关水利行业价格、税收、信贷、财务等政策，配合有关部门制定本省政策措施并组织实施；按照国家有关规定监督管理水利系统国有资产。 ——组织实施取水许可、水资源有偿使用、水资源论证等制度；组织水资源调查、评价和监测工作；组织编制水资源保护规划；指导饮用水水源保护；城市供水水源规划、城市防洪、城市污水处理回用等非常规水资源开发利用等；指导人河排污口设置工作；指导计划用水和节约用水工作。 ——承担水土流失综合防治工作；组织编制水土保持规划并监督实施；组织水土流失监测、预报并公告；审核大中型开发建设项目水土保持方案并监督实施。	水资源处、水保处

资料来源：根据河北省政府网站资料、调研情况整理。

附件（三）近年河北省财政预算管理改革的主要内容

1. 完善政府预算体系

（1）实行全口径预算管理。将省级政府所有收支都纳入预算管理，在明确收支范围的基础上，分别编制一般公共预算①、政府性基金预算②、国有资本经营预算③、社会保险基金预算④，建立定位清晰、分工明确的政府预算体系。政府性基金预算、国有资本经营预算、社会保险基金预算与一般公共预算相衔接（见图1）。

---

① 一般公共预算是对以税收为主体的财政收入，安排用于保障和改善民生、推动经济社会发展、维护国家安全、维持国家机构正常运转等方面的收支预算。

② 政府性基金预算是对依照法律、行政法规的规定在一定期限内向特定对象征收、收取或者以其他方式筹集的资金，专项用于特定公共事业发展的收支预算。从2015年1月1日起，将政府性基金预算中用于提供基本公共服务以及主要用于人员和机构运转等方面的项目收支转列一般公共预算，具体包括地方教育附加、文化事业建设费、残疾人就业保障金、从地方土地出让收益计提的农田水利建设和教育资金、转让政府还贷道路收费权收入、育林基金、森林植被恢复费、水利建设基金、船舶港务费、长江口航道维护收入等11项基金。

③ 国有资本经营预算是对国有资本收益作出支出安排的收支预算。国有资本经营预算按照收支平衡的原则编制，不列赤字，并安排资金调入一般公共预算。除调入一般公共预算和补充社保基金外，限定用于解决国有企业历史遗留问题及相关改革成本支出、对国有企业的资本金注入及国有企业政策性补贴等方面。一般公共预算安排的用于这方面的资金逐步退出。

④ 社会保险基金预算是对社会保险缴款、一般公共预算安排和其他方式筹集的资金，专项用于社会保险的收支预算。社会保险基金预算按照统筹层次和社会保险项目分别编制，做到收支平衡。

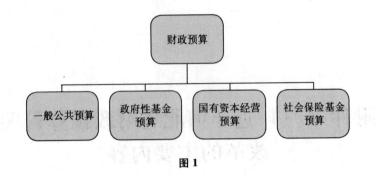

图 1

四本预算侧重有所不同，关于应对气候变化的财政支出包含在一般公共预算中。因此，在本报告分析中，主要针对一般公共预算中的相关支出进行分析。

（2）强化政府性基金、国有资本经营预算与一般公共预算的统筹。按照中央部署，逐步取消城市维护建设税、排污费、探矿权和采矿权价款、矿产资源补偿费等专款专用的规定，在一般公共预算中统筹安排相关经费。统一预算分配，逐步将所有预算资金纳入财政部门统筹配置。

（3）强化本级与上级资金统筹。河北省政府编制年初预算，将上级税收返还、下级上解收入、列入基数和提前通知的上级转移支付，与本级收入一并列入收入预算，统筹安排本级支出和对下转移支付。

（4）完善预算标准体系。河北省充分发挥支出标准在预算编制和管理中的基础支撑作用，健全基本支出定额标准体系，完善机关运行经费实物定额和服务标准，加快项目支出定额标准体系建设。建立定额标准动态调整机制，根据经济社会发展、政策变化适时进行调整。加强人员编制和资产管理，完善人员编制管理、资产管理与预算编制相结合的机制。建立部门预算基础信息库，夯实基本支出预算管理基础。

2. 改进预算控制方式

（1）实行中期财政规划管理

河北省从编制 2016 年预算开始，试编中期财政规划。根据经济运行情况、宏观调控方向，科学预测未来 3 年财政收入情况，全面梳理分析重大改革和支出政策，统筹编制本级 3 年滚动财政规划，并与本地经济社会发展规划纲要和国家宏观调控政策相衔接。年度预算编制与中期财政规划相衔接。各级各部门研究制定部门规划、行业规划，凡涉及财政政策和资金支持的，要与中期财政规划相衔接。加强预算项目库建设，健全项目申报审核机制，实现预算项目的滚动管理。

（2）建立跨年度预算平衡机制

河北省各级一般公共预算按规定设置预算稳定调节基金，用于弥补以后年度预算资金的不足。一般公共预算执行中如出现超收，用于化解政府债务或补充预算稳定调节基金；如出现短收，通过调入预算稳定调节基金或其他预算资金、削减支出实现平衡。如采取上述措施后仍不能实现平衡，省级经省人大或其常委会批准后增列赤字，并报财政部备案，在下一年度预算中予以弥补；市、县通过申请上级政府临时救助实现平衡，并在下一年度预算中归还。政府性基金预算和国有资本经营预算，如出现超收，结转下年安排；如出现短收，通过削减支出实现平衡。

3. 深化绩效预算改革

（1）全面推进绩效预算改革

《河北省人民政府关于深化绩效预算管理改革的意见》（冀政〔2014〕76号）要求，加快建立全过程绩效预算管理新机制。2015年在11个设区市本级和定州、辛集市全面实施，每个设区市再选择3个县（市、区）同步试点；2016年所有市、县（市、区）全面推开。

（2）改进预算审核方式

河北省各级财政部门审核预算，首先审核部门职责目标与政府工作的匹配性，再审核各项工作活动绩效目标指标的科学性，然后审核预算项目与职责活动的关联性、立项的必要性，最终合理确定项目预算额度，确保政府重大决策部署的全面落实，促进政府管理效能提升。

（3）全面推行绩效评价

河北省采取部门自评与财政评价相结合的方式，全面开展绩效评价。各部门负责"预算项目"层面的绩效评价，对年度完成情况全面自评；财政部门负责"工作活动"层面的绩效评价，并对重点领域、重大项目进行再评价。根据预算管理需要，拓展绩效评价范围，创新绩效评价方式，将绩效评价重点由项目支出，拓展到部门整体支出和政策、制度、管理等方面。

（4）加强评价结果应用

河北省建立预算绩效与预算安排挂钩机制，将绩效评价结果作为调整支出结构、完善财政政策和科学安排预算的重要依据。完善绩效评价报告制度和绩效问责制度，加大绩效信息公开力度。

4. 改进财政投入方式

(1) 加大政府向社会力量购买服务力度

将适合采取市场化方式提供、社会力量能够承担的公共服务，按照一定的方式和程序，交由具备条件的社会力量承担，并由政府根据服务数量和质量向其支付费用。所有适宜向社会力量购买的政府事务性管理服务，原则上都要引入竞争机制，通过合同、委托等方式向社会购买，纳入政府向社会力量购买服务的范围管理。

(2) 大力推广政府和社会资本合作模式（PPP模式）

河北省鼓励社会资本通过特许经营等方式，参与城市基础设施等有一定收益的公益性事业投资和运营，对价格调整机制相对灵活、市场化程度相对较高、投资规模相对较大、需求长期稳定的准公益类建设项目，探索运用规范的政府和社会资本合作模式，撬动社会资本参与公共产品供给。

(3) 实行评价后补助

对于政府支持鼓励的科技开发、科技服务等项目，由原来的事前补助资金，改为单位先行投入，取得成果或服务绩效后，由财政与有关部门验收审查或评价绩效，再给予补助，更好地发挥财政资金的引导作用。

(4) 积极推行股权投资

统筹政府支持产业发展的各类财政资金，设立产业引导股权投资基金，采取市场化方式运作，吸引社会资本支持经济结构调整和产业转型升级，形成财政手段和金融手段相配合的资金投入机制。

附件（四）河北省一般公共预算支出与气候的相关度分类分析表

科目编码	科目名称	高度相关	中度相关	低度相关
20104	发展与改革事务		√	
2010401	行政运行		√	
2010402	一般行政管理事务		√	
2010404	战略规划与实施		√	
2010408	物价管理		√	
2010450	事业运行		√	
2010499	其他发展与改革事务支出		√	
20105	统计信息事务			√
2010501	行政运行			√
2010505	专项统计业务			√
2010506	统计管理			√
2010507	专项普查活动			√
2010508	统计抽样调查			√
2010550	事业运行			√
20106	财政事务			√
2010601	行政运行			√
2010602	一般行政管理事务			√
2010603	机关服务			√
2010604	预算改革业务			√

续表

科目编码	科目名称	高度相关	中度相关	低度相关
2010605	财政国库业务			√
2010606	财政监察			√
2010650	事业运行			√
2010699	其他财政事务支出			√
2040299	其他公安支出			√
20601	科学技术管理事务			√
2060101	行政运行			√
2060102	一般行政管理事务			√
2060103	机关服务			√
2060199	其他科学技术管理事务支出			√
20602	基础研究		√	
2060201	机构运行		√	
2060203	自然科学基金		√	
2060204	重点实验室及相关设施		√	
2060206	专项基础科研		√	
2060299	其他基础研究支出		√	
20603	应用研究			√
2060301	机构运行			√
2060302	社会公益研究			√
2060399	其他应用研究支出			√
20604	技术研究与开发			√
2060401	机构运行			√
2060402	应用技术研究与开发			√
2060403	产业技术研究与开发			√
2060404	科技成果转化与扩散			√
2060499	其他技术研究与开发支出			√
20605	科技条件与服务			√
2060501	机构运行			√
2060503	科技条件专项			√
2060599	其他科技条件与服务支出			√

附件（四）河北省一般公共预算支出与气候的相关度分类分析表

续表

科目编码	科目名称	高度相关	中度相关	低度相关
20606	社会科学			√
2060601	社会科学研究机构			√
2060602	社会科学研究			√
20607	科学技术普及			√
2060701	机构运行			√
2060702	科普活动			√
2060703	青少年科技活动			√
2060704	学术交流活动			√
2060705	科技馆站			√
2060799	其他科学技术普及支出			√
20608	科技交流与合作			√
2060801	国际交流与合作			√
2060899	其他科技交流与合作支出			√
20699	其他科学技术支出			√
2069901	科技奖励			√
2069999	其他科学技术支出			√
21101	环境保护管理事务		√	
2110101	行政运行		√	
2110102	一般行政管理事务		√	
2110103	机关服务		√	
2110104	环境保护宣传		√	
2110199	其他环境保护管理事务支出		√	
21102	环境监测与监察	√		
2110203	建设项目环评审查与监督	√		
2110204	核与辐射安全监督	√		
21103	污染防治	√		
2110301	大气	√		
2110304	固体废弃物与化学品	√		
2110305	放射源和放射性废物监管	√		
2110307	排污费安排的支出	√		

续表

科目编码	科目名称	高度相关	中度相关	低度相关
2110399	其他污染防治支出	√		
21104	自然生态保护	√		
2110401	生态保护	√		
2110402	农村环境保护	√		
2110403	自然保护区	√		
21110	能源节约利用	√		
2111001	能源节约利用	√		
21111	污染减排	√		
2111101	环境监测与信息	√		
2111102	环境执法监察	√		
212	城乡社区支出			√
21201	城乡社区管理事务			√
2120101	行政运行			√
2120105	工程建设标准规范编制与监管			√
2120106	工程建设管理			√
2120107	市政公用行业市场监管			√
2120108	国家重点风景区规划与保护		√	
2120109	住宅建设与房地产市场监管			√
2120110	执业资格注册、资质审查			√
2120199	其他城乡社区管理事务支出			√
21202	城乡社区规划与管理		√	
2120201	城乡社区规划与管理		√	
21203	城乡社区公共设施		√	
2120399	其他城乡社区公共设施支出		√	
21205	城乡社区环境卫生		√	
2120501	城乡社区环境卫生		√	
21206	建设市场管理与监督			√
2120601	建设市场管理与监督			√
213	农林水支出			√
21301	农业			√

续表

科目编码	科目名称	高度相关	中度相关	低度相关
2130101	行政运行			√
2130102	一般行政管理事务			√
2130103	机关服务			√
2130104	事业运行			√
2130106	科技转化与推广服务			√
2130108	病虫害控制			√
2130109	农产品质量安全			√
2130110	执法监管			√
2130111	统计监测与信息服务			√
2130112	农业行业业务管理			√
2130119	防灾救灾	√		
2130122	农业生产资料与技术补贴			√
2130123	农业生产保险补贴			√
2130124	农业组织化与产业化经营			√
2130125	农产品加工与促销			√
2130126	农村公益事业			√
2130135	农业资源保护修复与利用	√		
2130152	对高校毕业生到基层任职补助			√
2130153	草原植被恢复费安排的支出			√
2130199	其他农业支出			√
21302	林业	√		
2130201	行政运行	√		
2130203	机关服务	√		
2130204	林业事业机构	√		
2130205	森林培育	√		
2130206	林业技术推广	√		
2130207	森林资源管理	√		
2130208	森林资源监测	√		
2130209	森林生态效益补偿	√		
2130211	动植物保护	√		

续表

科目编码	科目名称	高度相关	中度相关	低度相关
2130213	林业执法与监督	√		
2130216	林业检疫检测	√		
2130217	防沙治沙	√		
2130218	林业质量安全	√		
2130219	林业工程与项目管理	√		
2130221	林业产业化	√		
2130224	林业政策制定与宣传	√		
2130233	森林保险保费补贴	√		
2130234	林业防灾减灾	√		
2130299	其他林业支出	√		
21303	水利		√	
2130301	行政运行		√	
2130303	机关服务		√	
2130304	水利行业业务管理		√	
2130305	水利工程建设		√	
2130306	水利工程运行与维护		√	
2130309	水利执法监督		√	
2130310	水土保持		√	
2130311	水资源节约管理与保护		√	
2130313	水文测报		√	
2130314	防汛		√	
2130315	抗旱		√	
2130316	农田水利		√	
2130317	水利技术推广		√	
2130331	水资源费安排的支出		√	
2130399	其他水利支出		√	
21304	南水北调		√	
2130401	行政运行		√	
2130405	政策研究与信息管理		√	
2130407	前期工作		√	

附件（四） 河北省一般公共预算支出与气候的相关度分类分析表

续表

科目编码	科目名称	高度相关	中度相关	低度相关
2130408	南水北调技术推广		√	
2130499	其他南水北调支出		√	
21305	扶贫			√
2130501	行政运行			√
2130550	扶贫事业机构			√
2130599	其他扶贫支出			√
21306	农业综合开发			√
2130601	机构运行			√
2130602	土地治理			√
2130603	产业化经营			√
2130699	其他农业综合开发支出			√
21399	其他农林水支出			√
2139999	其他农林水支出			√
214	交通运输支出			√
21401	公路水路运输			√
2140101	行政运行			√
2140102	一般行政管理事务			√
2140104	公路新建			√
2140105	公路改建			√
2140106	公路养护			√
2140108	公路路政管理			√
2140112	公路运输管理			√
2140113	公路客货运站（场）建设			√
2140123	航道维护			√
2140127	船舶检验			√
2140131	海事管理			√
2140136	水路运输管理支出			√
2140199	其他公路水路运输支出			√
21402	铁路运输		√	
2140206	铁路安全		√	

续表

科目编码	科目名称	高度相关	中度相关	低度相关
2140208	行业监管		√	
2140299	其他铁路运输支出		√	
21403	民用航空运输			√
2140301	行政运行			√
2140302	一般行政管理事务			√
2140304	机场建设			√
21499	其他交通运输支出			√
2149999	其他交通运输支出			√
215	资源勘探信息等支出			√
21501	资源勘探开发			√
2150199	其他资源勘探业支出			√
21605	旅游业管理与服务支出			
2160501	行政运行			
2160502	一般行政管理事务			
2160504	旅游宣传			
2160505	旅游行业业务管理			
2160599	其他旅游业管理与服务支出			
21999	其他支出			
	国土海洋气象等支出			
	国土资源事务			√
2200101	行政运行			√
2200103	机关服务			√
2200107	国土资源社会公益服务			√
2200108	国土资源行业业务管理			√
2200113	地质及矿产资源调查			√
2200120	矿产资源专项收入安排的支出			√
2200150	事业运行			√
2200199	其他国土资源事务支出			√
22002	海洋管理事务			√
2200208	海洋执法监察			√
2200214	海域使用金支出			√

续表

科目编码	科目名称	高度相关	中度相关	低度相关
2200217	无居民海岛使用金支出			√
2200250	事业运行			√
22003	测绘事务			√
2200301	行政运行			√
2200304	基础测绘			√
2200305	航空摄影			√
2200350	事业运行			√
2200399	其他测绘事务支出			√
22004	地震事务			√
2200404	地震监测			√
2200405	地震预测预报			√
2200406	地震灾害预防	√		
2200407	地震应急救援	√		
2200450	地震事业机构			√
2200499	其他地震事务支出			√
22005	气象事务	√		
2200504	气象事业机构	√		
2200509	气象服务	√		
2200510	气象装备保障维护	√		
2200599	其他气象事务支出			√

附件（五）河北省一般公共预算支出根据减缓和适应气候变化相关度细分类表

表1　高度相关一般公共预算支出根据减缓和适应气候变化相关度细分类表

科目编码	科目名称	减缓			适应		
		高度相关	中度相关	低度相关	高度相关	中度相关	低度相关
21102	环境监测与监察	√			√		
2110203	建设项目环评审查与监督	√			√		
2110204	核与辐射安全监督	√			√		
21103	污染防治	√					
2110301	大气	√					
2110304	固体废弃物与化学品	√					
2110305	放射源和放射性废物监管	√					
2110307	排污费安排的支出	√					
2110399	其他污染防治支出	√					
21104	自然生态保护	√					
2110401	生态保护	√					
2110402	农村环境保护	√					
2110403	自然保护区	√					
21110	能源节约利用	√					
2111001	能源节约利用	√					
21111	污染减排	√					
2111101	环境监测与信息	√					

附件（五）河北省一般公共预算支出根据减缓和适应气候变化相关度细分类表

续表

科目编码	科目名称	减缓			适应		
		高度相关	中度相关	低度相关	高度相关	中度相关	低度相关
2111102	环境执法监察	√					
2130119	防灾救灾	√					
2130135	农业资源保护修复与利用	√					
21302	林业	√					
2130201	行政运行	√					
2130203	机关服务	√					
2130204	林业事业机构	√					
2130205	森林培育	√					
2130206	林业技术推广	√					
2130207	森林资源管理	√					
2130208	森林资源监测	√					
2130209	森林生态效益补偿	√					
2130211	动植物保护	√					
2130213	林业执法与监督	√					
2130216	林业检疫检测	√					
2130217	防沙治沙	√					
2130218	林业质量安全	√					
2130219	林业工程与项目管理	√					
2130221	林业产业化	√					
2130224	林业政策制定与宣传	√					
2130233	森林保险保费补贴	√					
2130234	林业防灾减灾	√					
2130299	其他林业支出	√					
2200406	地震灾害预防	√					
2200407	地震应急救援	√					
22005	气象事务	√					
2200504	气象事业机构	√					
2200509	气象服务	√					
2200510	气象装备保障维护	√					

表2　中度相关一般公共预算支出根据减缓和适应气候变化相关度细分类表

科目编码	科目名称	减缓			适应		
		高度相关	中度相关	低度相关	高度相关	中度相关	低度相关
20104	发展与改革事务		√			√	
2010401	行政运行		√			√	
2010402	一般行政管理事务		√			√	
2010404	战略规划与实施		√			√	
2010408	物价管理		√			√	
2010450	事业运行		√			√	
2010499	其他发展与改革事务支出		√			√	
20602	基础研究		√			√	
2060201	机构运行		√			√	
2060203	自然科学基金		√			√	
2060204	重点实验室及相关设施		√			√	
2060206	专项基础科研		√			√	
2060299	其他基础研究支出		√			√	
21101	环境保护管理事务		√			√	
2110101	行政运行		√			√	
2110102	一般行政管理事务		√			√	
2110103	机关服务		√			√	
2110104	环境保护宣传		√			√	
2110199	其他环境保护管理事务支出		√			√	
2120108	国家重点风景区规划与保护		√			√	
21202	城乡社区规划与管理					√	
2120201	城乡社区规划与管理					√	
21203	城乡社区公共设施					√	
2120399	其他城乡社区公共设施支出					√	
21205	城乡社区环境卫生					√	
2120501	城乡社区环境卫生					√	
21303	水利		√			√	
2130301	行政运行		√			√	
2130303	机关服务		√			√	
2130304	水利行业业务管理		√				
2130305	水利工程建设		√				

附件（五）河北省一般公共预算支出根据减缓和适应气候变化相关度细分类表

续表

科目编码	科目名称	减缓			适应		
		高度相关	中度相关	低度相关	高度相关	中度相关	低度相关
2130306	水利工程运行与维护		√				
2130309	水利执法监督		√				
2130310	水土保持		√				
2130311	水资源节约管理与保护		√				
2130313	水文测报		√				
2130314	防汛		√				
2130315	抗旱		√				
2130316	农田水利		√				
2130317	水利技术推广		√				
2130331	水资源费安排的支出		√				
2130399	其他水利支出		√				
21304	南水北调		√				
2130401	行政运行		√			√	
2130405	政策研究与信息管理		√			√	
2130407	前期工作		√			√	
2130408	南水北调技术推广		√				
2130499	其他南水北调支出		√				
21402	铁路运输		√				
2140206	铁路安全		√				
2140208	行业监管		√			√	
2140299	其他铁路运输支出		√				

表3 低度相关一般公共预算支出根据减缓和适应气候变化相关度细分类表

科目编码	科目名称	减缓			适应		
		高度相关	中度相关	低度相关	高度相关	中度相关	低度相关
20105	统计信息事务						√
2010501	行政运行						√
2010505	专项统计业务						√
2010506	统计管理						√
2010507	专项普查活动						√
2010508	统计抽样调查						√

续表

科目编码	科目名称	减缓			适应		
		高度相关	中度相关	低度相关	高度相关	中度相关	低度相关
2010550	事业运行						√
20106	财政事务						√
2010601	行政运行						√
2010602	一般行政管理事务						√
2010603	机关服务						√
2010604	预算改革业务						√
2010605	财政国库业务						√
2010606	财政监察						√
2010650	事业运行						√
2010699	其他财政事务支出						√
2040299	其他公安支出			√			√
20601	科学技术管理事务			√			√
2060101	行政运行			√			√
2060102	一般行政管理事务			√			√
2060103	机关服务			√			√
2060199	其他科学技术管理事务支出			√			√
20603	应用研究			√			√
2060301	机构运行			√			√
2060302	社会公益研究			√			
2060399	其他应用研究支出			√			
20604	技术研究与开发			√			
2060401	机构运行			√		√	
2060402	应用技术研究与开发			√			
2060403	产业技术研究与开发			√			
2060404	科技成果转化与扩散			√			
2060499	其他技术研究与开发支出			√			
20605	科技条件与服务			√			
2060501	机构运行			√			
2060503	科技条件专项			√			
2060599	其他科技条件与服务支出			√			
20606	社会科学			√			

附件（五）河北省一般公共预算支出根据减缓和适应气候变化相关度细分类表

续表

科目编码	科目名称	减缓			适应		
		高度相关	中度相关	低度相关	高度相关	中度相关	低度相关
2060601	社会科学研究机构			√			
2060602	社会科学研究			√			
20607	科学技术普及			√			
2060701	机构运行			√			
2060702	科普活动			√			
2060703	青少年科技活动			√			
2060704	学术交流活动			√			
2060705	科技馆站			√			
2060799	其他科学技术普及支出			√			
20608	科技交流与合作			√			
2060801	国际交流与合作			√			
2060899	其他科技交流与合作支出			√			
20699	其他科学技术支出			√			
2069901	科技奖励			√			
2069999	其他科学技术支出			√			
212	城乡社区支出						√
21201	城乡社区管理事务						√
2120101	行政运行		√				√
2120105	工程建设标准规范编制与监管		√				√
2120106	工程建设管理		√				√
2120107	市政公用行业市场监管			√			
2120109	住宅建设与房地产市场监管			√			
2120110	执业资格注册、资质审查			√			
2120199	其他城乡社区管理事务支出			√			
21206	建设市场管理与监督			√			
2120601	建设市场管理与监督			√			
213	农林水支出			√			
21301	农业			√			
2130101	行政运行			√			√
2130102	一般行政管理事务			√			√
2130103	机关服务			√			√

续表

科目编码	科目名称	减缓			适应		
		高度相关	中度相关	低度相关	高度相关	中度相关	低度相关
2130104	事业运行			√			√
2130106	科技转化与推广服务			√			√
2130108	病虫害控制			√			
2130109	农产品质量安全			√			√
2130110	执法监管			√			√
2130111	统计监测与信息服务			√			
2130112	农业行业业务管理			√			√
2130122	农业生产资料与技术补贴			√			
2130123	农业生产保险补贴			√			
2130124	农业组织化与产业化经营			√			
2130125	农产品加工与促销			√			
2130126	农村公益事业			√			
2130152	对高校毕业生到基层任职补助			√			
2130153	草原植被恢复费安排的支出			√			
2130199	其他农业支出			√			
21305	扶贫			√			
2130501	行政运行			√			
2130550	扶贫事业机构			√			
2130599	其他扶贫支出			√			
21306	农业综合开发			√			
2130601	机构运行			√			
2130602	土地治理			√			
2130603	产业化经营			√			
2130699	其他农业综合开发支出			√			
21399	其他农林水支出			√			
2139999	其他农林水支出			√			
214	交通运输支出			√			
21401	公路水路运输			√			
2140101	行政运行			√			√
2140102	一般行政管理事务			√			√
2140104	公路新建			√			√

附件（五）河北省一般公共预算支出根据减缓和适应气候变化相关度细分类表

续表

科目编码	科目名称	减缓			适应		
		高度相关	中度相关	低度相关	高度相关	中度相关	低度相关
2140105	公路改建			√			√
2140106	公路养护			√			
2140108	公路路政管理			√			
2140112	公路运输管理			√			
2140113	公路客货运站（场）建设			√			
2140123	航道维护			√			
2140127	船舶检验			√			
2140131	海事管理			√			
2140136	水路运输管理支出			√			
2140199	其他公路水路运输支出			√			
21403	民用航空运输			√			
2140301	行政运行			√			√
2140302	一般行政管理事务			√			√
2140304	机场建设			√			
21499	其他交通运输支出			√			
2149999	其他交通运输支出			√			
215	资源勘探信息等支出			√			
21501	资源勘探开发			√			
2150199	其他资源勘探业支出			√			
	国土资源事务			√			√
2200101	行政运行			√			√
2200103	机关服务			√			√
2200107	国土资源社会公益服务			√			
2200108	国土资源行业业务管理			√			
2200113	地质及矿产资源调查			√			
2200120	矿产资源专项收入安排的支出			√			
2200150	事业运行			√			√
2200199	其他国土资源事务支出			√			
22002	海洋管理事务			√			
2200208	海洋执法监察			√			
2200214	海域使用金支出			√			

续表

科目编码	科目名称	减缓			适应		
		高度相关	中度相关	低度相关	高度相关	中度相关	低度相关
2200217	无居民海岛使用金支出			√			
2200250	事业运行			√			
22003	测绘事务			√			
2200301	行政运行			√			√
2200304	基础测绘			√			√
2200305	航空摄影			√			√
2200350	事业运行			√			√
2200399	其他测绘事务支出			√			√
22004	地震事务			√			
2200404	地震监测			√			
2200405	地震预测预报			√			
2200450	地震事业机构			√			
2200499	其他地震事务支出			√			
2200599	其他气象事务支出			√			

参考文献

1. 中华人民共和国财政部：《2016年政府收支分类科目》，中国财政经济出版社，2015年。
2. 刘尚希："不确定性：财政改革面临的挑战"，《财政研究》，2015年第12期。
3. 刘尚希："关于实体经济企业降成本的看法"，《财政研究》，2016年第11期。
4. 刘尚希、石英华、罗宏毅："去产能不应是目标"，《财政研究简报》，2017年第12期。
5. 苏明、王桂娟等：《中国气候公共财政统计分析研究》，2015年3月，UNDP。
6. 苏明、石英华、王桂娟、陈新平："中国促进低碳经济发展的财政政策研究"，《财贸经济》，2011年第10期。
7. 傅志华、石英华、封北麟、于长革："'十三五'时期推动京津冀协同发展的主要任务"，《经济研究参考》，2015年第62期。
8. 任泽平、张庆昌："供给侧改革去产能的挑战、应对、风险与机遇"，《发展研究》，2016年第4期。
9. 韩国高："供给侧改革下我国去产能的现状、挑战与对策分析"，《科技促进发展》，2016年第5期。
10. 刘桂环、张彦敏、石英华："建设生态文明背景下完善生态保护补偿机制的建议"，《环境保护》，2015年第11期。
11. 石英华："按照治理现代化的要求构建多元化的生态补偿资金机制"，

《环境保护》，2016年第10期。

12. 石英华："积极稳妥推行中期财政规划管理"，《公共财政研究》，2015年1月。

13. Mark Miller, 2012. "Climate Public expenditure and Institutional Reviews (CPEIRs) in Asia Pacific Region——What We learnt", UNDP, www.aideffectiveness.org/Climate Change Finance.

14. Adelante, Hanh Le, 2015. "A methodological Guidebook, Climate Public expenditure and Institutional Review", UNDP.

15. Governance of Climate Change Finance Team, 2015 (UNDP Bangkok Regional Hub), "Budgeting for climate change: How governments have used national budgets to articulate a response to climate change, lessons learned from over twenty climate public expenditure and institutional reviews", UNDP.

16. Hanh Le, Kevork Baboyan, 2015 "The case study of: Bangladesh, Indonesia, Nepal and the Philippines", UNDP draft working paper.

17. Kit Nicholson, Thomas Beloe, Glenn Hodes, 2016. "Charting New Territory: A Stock Take of Climate Change Financing Frameworks in Asia - Pacific", UNDP.

Climate Public Expenditure and Institutional Review:

A Study of Hebei Province, China

Team Leader

LIU Shangxi, Director General, Chinese Academy of Fiscal Sciences

Coordinator

SHI Yinghua, Director, Center for Macroeconomic Studies, Chinese Academy of Fiscal Sciences

Members

PAN Liming, Ph. D. candidate, Chinese Academy of Fiscal Sciences

LUO Hongyi, Ph. D. candidate, Chinese Academy of Fiscal Sciences

Experts

GAO Zhili, Director General, Ministry of Finance, Hebei

YAO Shaoxue, Deputy Director General, Ministry of Finance, Hebei

LI Jiegang, Deputy Director General, Ministry of Finance, Hebei

LIU Qisheng, Director, Hebei Research Institute of Fiscal Sciences and Policies

ZHANG Shuo, Researcher, Hebei Research Institute of Fiscal Sciences and Policies

Reviewers

Thomas Beloe, Governance, Climate Change Finance and Development Effectiveness Advisor, Bangkok Regional Hub (BRH), UNDP Asia Pacific Regional Centre

Sujala Pant, Senior Advisor, Bangkok Regional Hub (BRH), UNDP Asia Pacific Regional Centre

Yuan Zheng, Economist, UNDP China

Foreword

The world is entering a new phase of development characterized by the Sustainable Development Goals (SDGs). The SDGs represent a higher level of ambition compared to the Millennium Development Goals (MDGs), given its attempt to achieve multi-dimensional sustainability. Environmental sustainability in particular, has gained increasing attention for its significance in ensuring sustained, inclusive and balanced economic growth. It requires a concerted effort to combat climate change and protect biodiversity as well as natural resources.

China has attached great importance to addressing climate change. It has enacted a series of goals to guide its efforts and investments. For instance, China submitted its Intended Nationally Determined Contributions (INDC) in 2016, putting forward its mitigation pledges in a domestic context. By 2030, China will aim to lower its carbon dioxide emissions per unit of GDP by 60% -65% from the 2005 level[①]. As China is the world's largest carbon-emitting country, it will also cap the carbon dioxide emissions by around the same time and make efforts to peak earlier. South-south cooperation on climate change is also an integral part of the INDC outline[②]. In line with the changing international landscape of global economic governance, China has recalibrated its position by scaling up the South-South Cooperation Fund on Climate Change (SSCFCC) to RMB20 billion (US $3.1 billion).

Expectations are therefore high for China to play a pioneering role in promoting green growth. The expectations are not unfounded. A lot has taken place in China at the top rungs of leadership and strategic development planning levels to prioritize green development, as evidenced in China's 13th FYP (2016 -2020).

---

① The climate change targets for China are highlighted in the INDC documents: Available here: http://www4.unfccc.int/submissions/INDC/Published% 20Documents/China/1/China's% 20INDC% 20 - % 20on% 2030% 20June% 202015. pdf.

② Available here: https://eneken.ieej.or.jp/data/6813.pdf.

1. Ecological civilisation in China's 13th FYP

The 13th FYP possesses five major development ideas: innovation, coordination, green, open and sharing. Green growth has, therefore, become one of the overarching principles to guide development by 2020. To accelerate green growth, China is in pursuit of an "Ecological Civilisation"; a new vanguard ideology highlighted in the 13th FYP. The concept was first proposed in the report of the 17th National Congress of Communist Party of China (CPC), and then integrated as one of the essential elements of the "five-in-one"; being given equal importance with the economic, political, cultural and social development outlined in the 18th National Congress.

The concept is rooted in Chinese philosophy, which states that humankind and nature are two inseparable parts and thus in unity. What defines ecological civilization has been intensively discussed and broadly understood from two perspectives[①]. First, seen from a temporal lens, an ecological civilization is considered a product evolved from industrialization. The idea is to address the balance between humankind and nature in the sense that societal development should rely on nature, making the best use of it while respecting, protecting and adapting to it. Second, seen from an endowment perspective, an ecological civilization is regarded as an overall framework that consists of several elements. This means, to construct an ecological civilization entails multi-dimensional efforts to adjust to patterns of production and consumption, institution set-ups, and the rule of law.

To sum up, an ecological civilization is the concept that "human prosperity can and should be achieved in a manner that respects the capacity of nature".[②]. It promotes the efficient use of natural resources and low-carbon development, as well as the safety and health of the natural environment. It is largely in line with the principles of sustainable development, which are the basis for the SDGs.

The 13th FYP has outlined a series of targets to be achieved to further green growth. These can be categorized into two main groups[③]. One group of targets

---

① Lian, G., 2014. Ecological civilization and green finance practice. China Finance Publishing House. (in Chinese)

② South-south cooperation for ecological civilization, 2016. China Council for International Cooperation on Environment and Development.

③ China's green growth roadmap in the 13th Five-Year Period, 2015. Global Green Growth Institute and Policy Research Center for Environment and Economy, Ministry of Environment Protection, China.

addresses relative green growth, focusing on the efficiency of resource/energy use, while the other group – absolute green growth – pays attention to total resource/energy consumption (Table 1).

Table 1   Main green growth targets outlined in the 13th FYP

Group	Targets
Relative green growth	– Energy consumption per unit of GDP to be reduced by 15% in 2020 (compared with 2015) – Reduction of carbon dioxide emissions per unit of GDP by 40% – 45% by 2020 (compared with 2015)[①].
Absolute green growth	– Reduction of water consumption by 35% by 2020 as compared with 2013[②] – Estimated total consumption of primary energy in 2020 of less than 5 billion tonnes of standard coal.

Moreover, a wide range of policy tools and measures are proposed to create an enabling environment for green growth. For instance, China will by 2020 establish an institutional framework composed of eight systems for promoting ecological progress, including a property rights system for natural resource assets and a system for developing and protecting territorial space[③]. To enhance environmental governance, China has also enacted the Environmental Protection Law and Air Pollution Control Law since 2015. In the 13th FYP period, the laws intend to impose a responsibility system to prevent improper interference from local governments and encourage public participation in mitigation obligations.

Another example relates to the projects planned for air, water and land pollution controls[④]. In terms of pollution control, there will be 10 million mu (6,667 square kilometers) of contaminated arable land restoration and 40 million mu (26,667 square kilometers) of contaminated farmland risk controls put in place to protect high-quality grasslands[⑤]. Regarding waste treatment measures, there will be the construction of five low-level radioactive waste disposal sites and a high radioactive waste disposal

---

① The target is consistent with China's Plan for Addressing Climate Change (2014 – 2020).

② The target is consistent with the target in the Water Pollution Prevention and Control Action.

③ The reform plan is published in full text. Available here: http://news.xinhuanet.com/english/china/2015-09/21/c_134646023.htm.

④ The full text of 13th FYP is released by NDRC. Available here: http://en.ndrc.gov.cn/newsrelease/201612/P020161207645765233498.pdf.

⑤ Source: Xinhua. Available here: http://en.xfafinance.com/html/Policy/2015/155187.shtml.

underground laboratory. To design green infrastructure and preserve natural capital, the government will set up a new soil erosion governance area of 270,000 square kilometres and a national wetland area of not less than 800 million mu (533,333 square kilometers). In total, 10 measures will be conducted for land pollution prevention, with 20 measures already taken for air and water pollution.

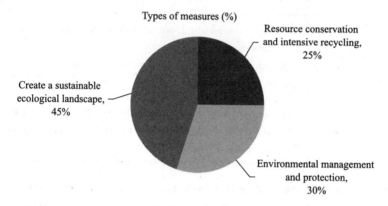

Figure 1　Purposes of green measures outlined in the 13th FYP

Source: 13th FYP.

2. Climate public expenditure institutional review

While more and more policy tools and measures are being put forward to advance green development in China, are there relevant institutional set-ups in place to make sure that policies are properly implemented? Has green growth been prioritized with enough monetary support? Have strategic green initiatives been budgeted for? Has green financing been aligned with green investment needs? All of these are key questions to address in order to enable evidence – based decision-making.

Against this backdrop, the Climate Public Expenditure Institutional Review (CPEIR) provides an opportunity to systematically review a country's institutional support, policy portfolios and public spending to see how they relate to climate-related activities. It has been initiated by the United Nations Development Program (UNDP) Asia-Pacific Regional Center and implemented in over 20 countries since 2011. The review serves as an innovative tool that ultimately helps to build a comprehensive climate change financing framework aimed at both national and sub – national levels. Its findings and recommendations strive to use climate change considerations in budget making and tagging, as well as support efforts that attempt to 'green-wash' the public financial management system.

In China's context, CPEIR hopes to help better understand the financial, institutional and policy issues that could lead to more effective changes at all pertinent levels, in order to further the ecological civilization vision. This exercise hopes to provide useful insights that could serve on-going efforts to promote green development, including green finance.

3. Green finance in China

China has invested heavily in support of the transition to greener development. Over the course of the 11th FYP (2006 – 2010), China invested approximately US $ 256 billion in the new energy sector and US $ 127 billion in energy efficiency[1]. In 2012, the central government allocated RMB97.9 billion (US$14.2 billion) to energy-saving emission reduction and renewable energy special funds, along with investment dollars for agriculture, water resources, marine management, health and meteorological activities[2].

However, the demand for financial support is significant in China, considering the scale of needed investment for sustainable energy, energy and resource efficiency, environmental remediation and protection and pollution control, as well as green infrastructure and products. In early 2015, the People's Bank of China estimated that the country needed annual investments of 3% of the country's GDP to address climate, water and land issues over the next five years[3].

In March 2016, the National People's Congress adopted the 13th FYP and explicitly proposed to "establish a green financial system and promote the development of green credit, green bonds and the establishment of green development funds."[4] Currently, green credit remains the primary source of green finance. By the end of 2014, China's 21 major banks registered in total about RMB6 trillion (US $ 1 trillion) in outstanding loans to green-credit-related customers, according to China Banking Regulatory Commission (CBRC)[5]. China became the largest green bond

---

[1] Available here: http://newenergynews.blogspot.tw/2013/03/todays-study-chinas-new-energy.html.

[2] Available here: http://www.sckxzx.com/index.php?_m = mod_article&_a = article_content&article_id = 98.

[3] Force, G. F. T. (2015). Establishing China's green financial system. People's Bank of China: Beijing.

[4] The original texts of 13th FYP are in Chinese. Available here: http://www.china.com.cn/lianghui/news/2016 – 03/17/content_38053101_13.htm.

[5] Green finance booming among Chinese banks. Available at: http://www.chinadaily.com.cn/bizchina/greenchina/2015 – 08/26/content_21709767.htm.

market in 2016, with the total of green bond issuance amounting to US $ 34 billion[①]. China's green equity market has also been developed with the disclosure of environmental information from listed companies and the launching of green securities indices and green investment funds. In addition, the insurance market has star ted to adopt Compulsory Environmental Liability Insurance (CELI) in China. The pilot results were, however, not fully satisfactory – with low participation of insurance companies and enterprises.

Another important vehicle to channel green finance is the Emissions Trading System (ETS) in China. The trading of carbon emission rights started running in seven locations in 2013: Beijing, Tianjin, Shanghai, Chongqing, Hubei, Guangdong and Shenzhen[②]. During this phase (2013 – 2015), the carbon market had a value of RMB1.1 billion (US $ 160 million) and remediated 68.6 megatons of carbon dioxide[③]. In a further sign of reform efforts, a nation-wide ETS will be launched in 2017 and actively engage with financial and intermediary institutions to develop innovative carbon financial products.

The People's Bank of China and seven other ministries and commissions officially released guidance for the construction of green financial systems in China on August 31, 2016[④]. Its goal is to open green finance to a full range of deployment avenues in China. Tax subsidies and commercial green bank subsidiaries are to be established based on the conventional green credit, bond, equity and insurance products[⑤]. Proposed incentives include offering a preferred deposit reserve ratio, releasing controls on supplementary mortgage loans, and allowing a less stringent loan risk weight. In addition, China will set up off-site carbon emissions derivatives businesses and promote environmental stress testing in financial institutions to strengthen international cooperation in the field of green finance continuously[⑥].

---

① Green finance progress report, 2017, UN Environment.

② Source: CCICED. Available here: http://www.cciced.net/ccicedEn/PublicationsDownload/201702/P020170210473297581978.pdf.

③ Source: The Climate Group. Available here: https://www.theclimategroup.org/news/spotlight-china-new-emissions-trading-system-set-revamp-global-market.

④ The document name is "Guidelines for Establishing the Green Financial System". Available here: http://gongwen.cnrencai.com/yijian/92291.html.

⑤ Green bank examples include Industrial Bank, CITIC Bank and ICBC. Available here: http://news.163.com/16/0229/02/BGV7K4LA00014AED.html.

⑥ Available here: http://www.nbd.com.cn/articles/2017-01-12/1069585.html.

4. The Belt and Road Initiative and green development

The Belt and Road Initiative (BRI) was initiated in China back in 2013. It aims to optimize resource allocation and utilization across countries and regions through connectivity in economy, finance, policy and culture, which has the potential to ultimately drive another wave of regional growth through international dialogue and cooperation.

The BRI has received a lot of attention globally given its great perceived potential to promote sustainable development. The initiative is found closely linked to the SDGs in multiple dimensions[①]. If the two can be effectively aligned and implemented, the BRI can help countries achieve their own development objectives within the space of sustainability.

In terms of green development, the BRI possesses lots of opportunities. It can be realized through varied means, including green infrastructure and green buildings. There are many more options. Green investments, green trade and green technology transfer can all broaden the scope of south-south cooperation in green development. Moreover, the establishment of a green financial system throughout BRI countries can act as another engine of green growth. Therefore, it is of strong interest to further investigate how to maximize the potential of the BRI to facilitate green growth.

5. The rationale, aims and objectives of the report

United Nations Development Program (UNDP), in collaboration with the Research Institute for Fiscal Sciences (RIFS, now renamed as Chinese Academy of Fiscal Science, CAFS), conducted the first Climate Public Expenditure Institutional Review (CPEIR) in China from 2014 to 2015. The Phase 1 report provides quantitative information on climate-related expenditures at the central government level. The research reviews a key national institutional framework that is critical to tackling climate change and effectively governing financial resource allocation. The study also touched upon climate governance experience in other developing countries and summarised lessons which may yield useful insights for China.

While China is achieving some success, the report documented the downward trend of its public expenditure. The total share of national government spending for energy conservation and environmental protection programs dropped from 2.7% in

---

[①] Pingfan Hong, 2016, "Jointly building the 'Belt and Road' towards the Sustainable Development Goals", United Nations Department of Economic and Social Affairs.

2010 to 1.3% in 2013. Adding to this, the report found that 7% of the central government budget was spent on measures directly or partially targeted to address climate change in 2014 (Figure 2).

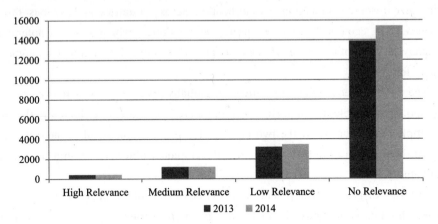

Figure 2　Central government expenditure classified according to climate relevance (Unit: 100 million RMB)

Source: Report on Climate Public Expenditure and Institutional Review in China.

Building on the success of the first phase of national CPEIR, UNDP and CAFS jointly initiated the second phase, where analyses will take place at the provincial level. Ultimately, central policies need to be localised to take effect. To achieve this, fiscal spending is of paramount significance to support and direct policy making. Therefore, a provincial CPEIR could help better understand the governance structure for climate action and the scale of climate-relevant spending at the provincial level, so as to provide useful information for budget planning. Going one step further, the report will preliminarily assess the cost-effectiveness of climate spending. This exercise is to help identify entry points where money could be better managed and spent to maximise expected impacts.

Specifically, the report aims to: (1) establish a baseline of climate-related expenditure at the provincial level; (2) assess whether or to what extent public expenditure on climate-related activities complies with policy priorities and strategies on climate change or green growth at the provincial level; (3) identify bottlenecks and opportunities to strengthen the enabling environment to attain the vision of these goals and policies.

UNDP
July 31st, 2017

Preface

Climate change issues have become increasingly serious, and tackling climate change is a task faced by the international community. China has been adopting an ecological civilization construction strategy and has actively participated in the global movement to combat climate change and promote sustainable development. In 2016, the Chinese Academy of Fiscal Sciences (CAFS) undertook the second phase of its Climate Public Expenditure and Institutional Review (CPEIR). This project, as of CAFS's completion of the first phase from 2014 to 2015, is also funded by the United Nations Development Program (UNDP). In the first phase, CAFS experts established a statistical methodology for China's climate public finance, and made a statistical analysis of the climate public expenditure incurred directly by the central government. The research achievements are highly recognized by the relevant institutions of the United Nations, the relevant divisions of Ministry of Finance and the National Development and Reform Commission (NDRC). The second phase will be advanced in depth, and will prioritize China's climate public expenditure and institutional review at the local level. By using case studies, experts will launch the provincial climate public expenditure and institutional review and explore the climate expenditure cost-benefit analysis approach.

CAFS is responsible for this research project and attaches great importance to the study of climate public expenditure. It has set up a research group headed by Liu Shangxi, Director General (DG) of CAFS, who has been selected as a member of the third National Climate Change Expert Committee (NCCEC). Gao Zhili, DG, Yao Shaoxue, Deputy Director General (DDG), Li Jiegang, DDG, Liu Qisheng, Director and Zhang Shuo, Associate Researcher of the Department of Finance in Hebei are invited as experts. The tasks of this project were preliminarily clarified in July 2016. Specifically, Hebei will be taken as the sample in the provincial CPEIR, and typical overcapacity reduction cases will be used in the cost-benefit analysis. In the course of the project, the project team conducted in-depth literature review and data collection, held a project kick-off meeting, went to Hebei two times to conduct intensive surveys, and regularly held discussions with UNDP and other international experts.

On October 10, 2016, the project team held a "Provincial Climate Public Expenditure and Institutional Review" kick-off meeting in Shijiazhuang City, Hebei. Among those present were experts from the Department of Finance (DOF), Department of Water Resources (DWR), Department of Forestry (DF), Health and Family Planning Commission (HFPC), Department of Environmental Protection (DEP), Development and Reform Commission (DRC), Department of Industry and Information Technology (DIIT), Department of Agriculture (DOA), Bureau of Meteorology (BOM) of Hebei, experts from the CAFS research group, officials from UNDP China. Liu Shangxi, DG of CAFS and research project leader, Yao Shaoxue, DDG of the DOF of Hebei, and Dr. Zheng Yuan, Economist UNDP China, addressed the meeting. Researcher Shi Yinghua introduced the research background, research approach and the proposed methodology on behalf of the research group to the audience. The experts present shared their views and opinions.

Liu Shangxi pointed out that the project is to assess fiscal expenditure's impact on mitigation and adaptation to climate change. Climate public expenditure is vital to the optimization of the fiscal expenditure structure. Expenditure is optimized only when it is evolving in favor of mitigation and adaptation to climate change. Climate change has a great impact on the survival and development of mankind and research in this regard is of great value. China is obliged to explore a new path to deal with climate change. This is what China should do as a major responsible country. If the research is carried out on a national scale, the workload will be heavy. Hebei Province, as an important member in the Beijing-Tianjin-Hebei region, is bold to be carrying out system reform and holds a leading position in combating climate change. Therefore, taking Hebei as the example in our research will help generate larger-scale effects. UNDP has extended great support to this project and deems it reasonable to target Hebei as a case study. This study is going to sum up experience from the concept, policy and institutional mechanism perspectives. Experts should not be content with existing practices, but need to continuously learn from emerging cases to thereby build up a model for developing countries.

At the first research seminar, the project team exchanged views with experts from authorities directly under provincial jurisdiction, including the DWR, DOF, HFPC,

Preface

DEP, DRC, DIIT, DOA and BOM, and held in-depth discussions on climate public expenditure and budget systems with experts from the functional divisions of the DOF of Hebei, including the Bureau of Management and Budget, Economic Construction Division, Division of Resource Conservation and Environmental Protection, Agriculture Division, General Office, Taxation Policy Division, Social Security Division, Public-Private-Partnership Office, Procurement Office, and Scientific Research Center.

Given that overcapacity reduction has been one of the important initiatives taken by Hebei to curb climate change in recent years, the project team used some overcapacity reduction cases in Hebei for research. From April 24 to 26, 2017, the project team, led by DG Liu Shangxi, went to Hebei for case research related to climate public expenditure cost-benefit analysis; Li Jiegang, deputy chief of the DOF of Hebei, provided guidance throughout the whole process. During the investigation, the project team had heated discussions with leaders and experts from the DRC, DOF, DIIT, DEP, DHRSS, and other authorities directly under the jurisdiction of Hebei; the provincial industrial associations, including the Metallurgical Association and the Coal Association; the BIIT of Shijiazhuang City and its subordinate Pingshan County; DRC, BOF, BIIT, BHRSS, BEP and other authorities directly under jurisdiction of Tangshan City; as well as the DRC, BOF, BHRSS, Government Office, BIIT directly under jurisdiction of Fengnan District; the team also visited a cement enterprise in Pingshan County, Tangshan Iron & Steel Group Co., Ltd. the largest subsidiaries of HBIS Group Co., Ltd., and Bainite Steel Group Co., Ltd., to explore the status quo of overcapacity reduction in Hebei, and the benefits, the challenges and problems thereof.

From September 5 to 9, 2016, the team members, including Shi Yinghua, Researcher and Director of the Macroeconomic Research Center, CAFS, Zhang Shuo, Associate Researcher from the Hebei research institute of Fiscal Science, and Dr. Zheng Yuan, Economist UNDP China went to Pakistan for a bilateral exchange meeting, "Integrating Climate Change Finance in Planning and Budgeting Systems". During their stay in Pakistan, the project team held discussions with Pakistan's top think tank, Leadership for Environment and Development (LEAD), on CPEIR, involving its goals, elements, methods, data acquisition, measurement index and time sensitivity, and met with Mr. Pated Ghazanfar Abbas Jilani, Assistant Minister of the Ministry of Finance of Pakistan, to learn about the progress of the government's CPEIR application, and Ms. Sujala Pant, Senior Advisor, Bangkok Regional Hub,

UNDP Asia Pacific Regional Centre. The team also visited the UNDP Pakistan, Ministry of Climate Change, General Accounting Department, Department of International Development of Pakistan, and participated in the CPEIR kick-off meeting held by LEAD in Lahore, Punjab. This bilateral exchange would help to deepen the understanding and consensus on the CPEIR-related approaches, the application of cost-effectiveness and in climate expenditure, thereby accelerating the progress of China's climate public expenditure review.

The research has yielded good results. Team members have been invited to the relevant climate change public expenditure policy research and consultation workshops as experts. On September 30, 2016, the project leader Liu Shangxi, a member of the 3rd NCCEC, participated in the inaugurating meeting and the first working meeting of the 3rd NCCEC. The project team has completed two research reports, namely, "China's Provincial Climate Public Expenditure and Institutional Review – Taking Hebei as an Example", and "Cost-Benefit Analysis of Climate Public Expenditure Project – Taking Hebei's Overcapacity Reduction Cases as an Example". Investigation achievements such as "Overcapacity Reduction is not the Ultimate Goal" and "Further Analysis of the Relation Between Overcapacity Reduction and Cost Reduction, De-leveraging and Increased Profits" have been submitted to the relevant government decision-making authorities in the form of briefing and special reports, and have received lot of attention from Xiao Jie, Minister of Finance.

On Sept. 8, 2017, Chinese Academy of Fiscal Sciences (CAFS) together with the United Nations Development Programme (UNDP) held a conference to launch the report entitled *Climate Public Expenditure and Institutional Review (CPEIR): A Study of Hebei Province, China*. Highly complimentary remarks were given by representatives including, Agi Veres (Country Director, UNDP China); Asad Maken (Senior Regional Advisor, UNDP Asia-Pacific Regional Center); Bie Tao (Director General Department of Policies, Laws and Regulations, Ministry of Environmental Protection); Zhang Xiaode (Professor, Director, Ecological Civilization Research Center, Chinese Academy of Governance); Li Jiegang (Deputy Director General, Department of Finance, Hebei Province); Zhu Shouqin (Senior Researcher, World Resources Institute); Huang Wenhang (Director, General Office, Department of Addressing Climate Change, National Development and Reform Commission); Guo Fengyuan (Deputy Director, General Office, Department of Energy Saving and Comprehensive Utilization, Ministry of Industry and Information Technology); Liu Zhe (Associate Researcher, Policy Research Center for Environment and Economy,

Preface

Ministry of Environmental Protection)。

During the investigation period, the project team received strong support and guidance from the authorities, industry associations, enterprises of Hebei. Leaders and experts of the DOF of Hebei also have made valuable contributions. The project team would like to thank the authorities and industry associations of Hebei, the Bureau of Industry and Information Technology of Shijiazhuang City and its subordinate Pingshan County, and the authorities of Tangshan City and their subordinate Fengnan District. The project team also would like to thank the enterprises involved in the field visits. Further, the support and guidance from the UNDP experts Niels Knudsen and Carsten Germer were indispensable for the achievements. Many thanks also go to UNDP's research assistants Wei Li and Andrew Cheng for their significant contribution to the project.

Climate public expenditure review is an emerging field. The statistical caliber and scope of climate public expenditure has not yet been defined, and the evaluation method remains to be improved. That's why climate public expenditure review is challenging. Though the study is still in its initial stage, the project team will continue to explore better approaches for curbing climate change and seek to make humble contributions to the betterment of mankind.

CAFS Research team
Sept. 30th, 2017

Executive Summary

Climate change has increasingly become a global challenge for sustainable development. Accelerating the building of an ecological system has been made a main measure by China to address climate change and maintain global ecological security. It has also been incorporated into top national social-economic development planning. Local governments have formulated many policies and regulations to address climate change, and budgeted for relevant activities. Provincial governments play an important role in this regard. The study of climate public expenditure and institutions at the provincial level is of great significance, as it portrays the dynamic practices at work in the establishment of an ecological civilization, explores a unique way of developing such a civilization, contributes to decision-making focused on addressing climate change and helps with the comparison of and exchanges between local climate public expenditure and that of other countries.

Hebei is located in the North China Plain, bordering Beijing and Tianjin with Bohai Bay to its east. Hebei is an industrial province. It has a large amount of steel production, accounting for 1/4 of China's total; it is thus burdened by the conflict between the use of environmental resources for economic development and the pressure and challenge of environmental governance. In recent years, Hebei has galvanized to combat climate change, witnessing an increase in climate – related public expenditure. Hebei has thus been chosen as a case to study the climate public expenditure of the local government, to understand where the local government's public expenditure is going, and to offer experience to, and promote the ongoing efforts of, other local governments.

I. Hebei climate public expenditure and institutional review

By evaluating Hebei's climate public expenditure and institutions, we reached the following conclusions:

1. Climate change has been given increasing attention, and climate-related public expenditure is growing year by year. Evaluation results showed that climate public expenditure increased year by year during the 12th FYP period. In particular, since 2015, Hebei has made more investment in energy conservation and environmental protection, and the public expenditure highly related to climate increased 38.7% compared to the previous year. Based on relevant analysis, public expenditure highly or moderately related to climate change increased from 8.23% of the total expenditure in 2011 to 11.29% in 2015. In terms of the share of climate spending in total fiscal expenditure, Hebei's climate public expenditure accounted for a higher percentage point than that of the central government.

2. The methodology to systematically review climate spending has been expanded. The first phase in the evaluation of climate public expenditure and its system was based on a statistical classification methodology of relevance to climate change, used to study expenditures at the central government level. In the second phase of the evaluation, more approaches were used to derive the desired statistics. Based on the statistical classification methodology by relevance, mitigation and adaptation to climate change were two core dimensions. This report found that statistics based on these dimensions were able to calculate climate public expenditure more precisely, which provided a foundation for cost-effective analysis and performance evaluation.

In response to the above-mentioned conclusions from the evaluation of climate public expenditure and its system, this report brings forward the following suggestions:

1. The vision and policies of top leadership can be further improved to form a sound interaction between the government, the market and the society.

2. China's budget management system can be further reformed to enhance the efficiency and effectiveness of policies and fund use.

3. Fiscal spending can be further refined, and a restraint mechanism controlled by the market can be established.

4. Research on the cost-benefit evaluation of public expenditures concerning climate-related issues can be strengthened.

II. Cost-benefit analysis of over-capacity reduction in Hebei

Climate change is closely related to economic development patterns. Turning toward a low-carbon model of economic development is necessary in order to address

Executive Summary

climate change issues. Currently, China is promoting supply-side structural reform, of which reducing overcapacity is one important aspect. By overcapacity reduction, production patterns can be optimized and upgraded, resource utilization efficiency could increase, industrial emission could be reduced, and the quality of economic development could be improved—it is a significant measure to cope with climate change. Considering that overcapacity reduction is one of the most important measures taken by Hebei in recent years, this report chose it as the case study. Through field investigations into relevant government agencies and enterprises, this report conducted a systematic analysis of the costs and benefits of overcapacity reduction in Hebei, and tried to establish a methodological framework for cost-benefit evaluation of public climate expenditure.

Through cost-benefit analysis of overcapacity reduction, this report has drawn the following conclusions:

1. Overcapacity reduction is a significant measure for dealing with climate change; more research should be done in this area. The promotion of low-carbon models of economic development is critical for China to address its domestic climate change-related challenges. Overcapacity reduction is necessary to close outdated production facilities and promote the transformation of economic development patterns, improve the quality of economic development and cope with climate change.

2. The high cost of overcapacity reduction is shared by relevant parties, including government and enterprises. Based on the features of the different parties involved, as well as on the concept and classification of visible cost, invisible cost and opportunity cost, this report presents the cost of overcapacity reduction in a matrix. This multi-dimensional matrix shows that in the process of overcapacity reduction, all parties, including government, enterprises and banks, pay very high prices. In the analysis of total cost, we should pay attention not only to the explicit expenditure of special funds provided by the central government, provincial government and county government, but also to the implicit expenditures spent by other related entities, including other government levels, enterprises and banks. Although it is hard to make an accurate quantitative analysis of the latter, implicit expenditures should not be ignored.

3. Overcapacity reduction has achieved notable results in terms of economic and social outcomes, as well as in the field of sustainable development. Overcapacity reduction by altering modes of production in the industrial process, aims at transforming China from an extensive economy to a circular economy by reducing the

pollution caused by industrial waste and by lowering the human impact on climate change. It is an integral part of constructing an ecological civilization; achieving such a civilization is critical for improving people's living standards and protecting the natural environment. Overcapacity reduction has been associated with social benefits, such as improving population health through the creation of a cleaner and less toxic environment. With a more responsible utilization of resources, sustainable development can be achieved, which is crucial for stabilizing employment, increasing worker incomes, and ultimately giving people a better livelihood. It also results in economic benefits, such as enhancing the quality of economic development, promoting the transformation and upgrading of the economy, and optimizing industrial structures and distribution channels.

4. In the short run, the cost of overcapacity reduction outweighs the benefits; in the long run, however, the benefits are greater than the cost. Combining field research of the enterprises in Hebei with the analysis of costs and benefits, this report concludes that the cost of overcapacity reduction is higher than the benefits in the short term, creating a stress on expenditures for the government, enterprises and banks over the present period. According to the data analysis of industrial liabilities, costs, and profits in Hebei, overcapacity reduction leads to the rise of costs and the decline of profit margins in the enterprises studied. However, in the long term, the benefits exceed the cost. From a microscopic perspective, overcapacity reduction ultimately contributes to cost reductions and an increase in profits. Overcapacity reduction, which is realized by improving production factors, including environmental protection, energy consumption, quality and safety, is able to eliminate outdated capacity effectively and lower the overall social production cost. With the adjustment of production structures, value is added and the profits of enterprises go up. From a macroscopic view, overcapacity reduction is beneficial for economic, social and ecological sustainability.

Based on our cost-benefit analysis of overcapacity reduction projects, this report brings forward the following suggestions:

1. Overcapacity reduction should be taken into consideration while developing strategies for achieving a circular economy and optimizing industrial distribution.

2. The policies related to the cutting of overcapacity should be implemented with corresponding tactics and dynamic optimization.

3. The policy of awards and subsidies implemented by the central government should be improved.

Executive Summary

4. The multi-dimensional matrix analysis framework of the costs and benefits of climate public expenditure needs further improvement.

5. The collection and consolidation of basic public climate expenditure related data needs further strengthening.

List of Acronyms

AAA	American Accounting Association
BEP	Bureau of Environmental Protection
BHRSS	Bureau of Human Resources and Social Security
BIIT	Bureau of Industry and Information Technology
BOF	Bureau of Finance
BOM	Bureau of Meteorology
CAFS	Chinese Academy of Fiscal Science
CBRC	China Banking Regulatory Commission
CCA	China Cost Association
CELI	Compulsory Environmental Liability Insurance
CPC	Communist Party of China
CPEIR	Climate Public Expenditure Institutional Review
DDG	Deputy Director General
DEP	Department of Environmental Protection
DF	Department of Forestry
DG	Director General
DIIT	Department of Industry and Information Technology
DOA	Department of Agriculture
DOF	Department of Finance
DRC	Development and Reform Commission
DWR	Department of Water Resources
ETS	Emissions Trading System
FYP	Five-Year Plan
GDP	Gross Domestic Product
HFPC	Health and Family Planning Commission
IPCC	Intergovernmental Panel on Climate Change

INDC	Intended Nationally Determined Contributions
LTFTCC	Leading Task Force on Tackling Climate Change in Hebei
MDGs	Millennium Development Goals
SDGs	Sustainable Development Goals
SSCF	South-South Cooperation Fund
UNDP	United Nations Development Program
UNFCCC	United Nations Framework Convention on Climate Change

Table of Content ▪ ▪ ▪ ▪

Report I: Climate Public Expenditure and Institutional Review at the Provincial Level in China ——A Study of Hebei Province (141)
  1. The necessity of climate public expenditure and institutional reviews at the local level (141)
    1.1 Provincial governments play a significant role in tackling climate change (142)
    1.2 Public expenditure is an important tool for provincial governments to address climate change (143)
    1.3 The importance of climate public expenditure and institutional review at the provincial level (146)
  2. Climate change and public policies in Hebei (147)
    2.1 The status of climate change in Hebei (147)
    2.2 Planning and regulation (149)
    2.3 Climate change institutional set-up in Hebei (155)
  3. Analysis of the budget system of climate public expenditure in Hebei (157)
    3.1 The structure of budget management (157)
    3.2 Budget preparation (157)
    3.3 Budget item setting (158)
  4. Statistics and assessment of climate public expenditure in Hebei (160)
    4.1 The determination of the scope of climate public expenditure in Hebei (160)
    4.2 Statistics on climate public expenditure in Hebei (161)
  5. Key conclusions and recommendations (171)
    5.1 Key conclusions (171)
    5.2 Recommendations (173)

Report II: Cost-benefit Analysis of Climate Public Expenditure ——A Case Study on Overcapacity Reduction in Hebei Province (176)
  Introduction (176)

1. Progress made in overcapacity reduction in Hebei ……………… (177)
   1.1 General information about overcapacity reduction in industries in Hebei …………………………………………………………… (177)
   1.2 Current developments in overcapacity reduction in Hebei …… (178)
   1.3 Future plan for overcapacity reduction in Hebei ……………… (179)
2. Analysis of cost of overcapacity reduction ……………………………… (180)
   2.1 Theoretical framework for cost analysis of overcapacity reduction ……………………………………………………………………… (180)
   2.2 Analysis of the government's cost of overcapacity reduction … (186)
   2.3 Analysis of enterprises' cost of overcapacity reduction ………… (192)
   2.4 Analysis of cost of overcapacity reduction by the financial system (with banks as an example) ……………………………………… (195)
3. Analysis of the benefits of overcapacity reduction …………………… (196)
   3.1 Eco-environmental benefits ……………………………………… (197)
   3.2 Social benefits analysis …………………………………………… (200)
   3.3 Economic benefits analysis ……………………………………… (201)
4. Key conclusions and recommendations ………………………………… (206)
   4.1 Key conclusions …………………………………………………… (206)
   4.2 Recommendations ………………………………………………… (207)

Conclusion ……………………………………………………………………… (210)
Annex I. Policies on climate public expenditure …………………………… (214)
Annex II. Related government bodies ……………………………………… (225)
Annex III. The main contents of budget management reform in Hebei ……………………………………………………………………………… (231)
Annex IV. Correlation between the classification of general public budget expenditure items and climate (detailed) ……………………… (236)
Annex V. Detailed items of general budget expenditure classified based on correlation with climate change mitigation and adaptation …… (246)
References ……………………………………………………………………… (259)

Report I:
Climate Public Expenditure and Institutional Review at the Provincial Level in China

——A Study of Hebei Province

1. The necessity of climate public expenditure and institutional review at the local level

Climate change and its impacts have raised worldwide attention. The Fifth Assessment Report of the United Nations Intergovernmental Panel on Climate Change (IPCC) noted that between 2000 and 2010, anthropogenic greenhouse gas emissions increased by an average of 2.2% per annum, higher than the average annual growth rate of 1.3% during the previous 30 years. Since start of the 21st century, the global sea level has risen 19cm, with a mean annual increase of 1.7mm. The human environment is changing at an alarming rate.

Increasing temperature, rising sea level, extreme climate and many other events pose a serious challenge to human survival and development. On the one hand, the IPCC report pointed out that since the beginning of this century, economic losses from natural disasters induced by global warming have been as high as US $ 2.5 trillion. By 2050, annual losses are expected to exceed US $ 1 trillion. Climate change also inflicts direct harm on people's health through heat stress response, the accelerated transmission of infectious diseases, and the deterioration of human living conditions.

China will accelerate the construction of an ecological civilization as a major measure to adapt to climate change and safeguard global ecological security; to do so, it has incorporated climate change into its national economic and social development planning. Furthermore, local governments have also developed a significant number of policies and regulations to cope with climate change effectively. Public expenditure is an important tool for local authorities to deal with climate change, and it is seen an essential part of evaluating the mechanisms for tackling climate change at the provincial level.

1.1 Provincial governments play a significant role in tackling climate change

1.1.1 Tackling climate change needs to be implemented through provincial governments

From the academic point of view, local implementation could help alleviate the problem of information asymmetry. Although central decision-making could overcome the externalities of tackling climate change, maintain a unified market and avoid the 'prisoner's dilemma', while taking into account information asymmetry, local governments have more informational advantages at the operational level than at the level of the central government. It could contribute to cost reduction if specific work is implemented locally.

From the perspective of current resource allocation, local regions are the main receivers. Take the four areas highly related to climate change, namely (a) energy conservation and environmental protection, (b) agriculture, forest and water resources, (c) land resources and marine meteorology and (d) heath care; the total national expenditure in 2015 was RMB3,625.126 billion (US$566.426 billion), among which central government expenditure was RMB151.764 billion (US$23.713 billion), accounting for only 4.34% of the national expenditure. In contrast, local expenditure accounted for more than 95%, assuming a role of vital importance.

1.1.2 Proper tackling climate change is the key to local sustainable development

On the one hand, climate and environmental issues have severely affected local economic and social development. Climate and environmental change increase the frequency and destructive power of extreme weather. Waterlogging, flood, drought, typhoon, hail, low temperature and frost weather, snow and other climatic conditions have seriously affected people's livelihoods. According to the data of the National Disaster Reduction Office, natural disasters in 2015 affected 186 million people in China and caused an economic loss of RMB270.4 billion (US$42.25 billion). Climate and environmental change have become the most pressing threat to local sustainable development. Local governments must properly adapt to climate change and achieve green low-carbon sustainable development.

On the other hand, proactive tackling climate change provides major opportunities to promote local economic and social development and transformation. China, a country with relatively poor resources, low per capita income and a large quantity of people in poverty, is facing many challenges, such as economic development, poverty reduction, livelihood improvement, environmental protection and tackling climate change. With the previous development model, the resources and energy needed for production and consumption in China have exceeded the ecological carrying power, resulting in an increase of pressure on the available resources and the environment as a whole. Active climate change adaption is needed to improve resource conservation, eliminate the pressure on resources and the environment and meet the demands for economic growth among those consumers with the least resources.

1.2 Public expenditure is an important tool for provincial governments to address climate change

1.2.1 Climate externalities requires financial support

It is generally believed that climate change refers to the rise in temperature and increase in storm activities at the global level, a statistically significant change in the average distribution of weather patterns or a change lasting for an extended period of time. Though climate change may be caused by an internal factor in nature, existing studies have shown that long-term climate change mainly results from human activities, such as carbon dioxide emissions. In Article 1 of the United Nations Framework Convention on Climate Change (UNFCCC), "climate change" is defined as "a change of climate which is attributed directly or indirectly to human activity that alters the composition of the global atmosphere and which is in addition to natural climate variability observed over comparable time periods."

Climate change is both an environmental issue and a development issue because climate change has distinct externalities that need to be managed. The existence of externalities leads to inconsistencies between private and social marginal costs and between private and social marginal revenue, which distorts the pricing signals and makes the market equilibrium of products and services based on market competition, as opposed to Pareto-optimal. Positive and negative externalities in climate change affect the efficiency of resource allocation and can lead to market failure. Market failure requires government intervention and financial support[①].

1.2.2 The public nature of finance determines the need to tackling climate change

In its 1997 World Development Report, the World Bank summarized that the core mission of each government consisted of five essential responsibilities, reflecting the general functions exercised by modern governments: determining a legal basis; maintaining an intact policy environment, including the maintenance of macroeconomic stability; investing in basic social services and infrastructure; protecting vulnerable groups; protecting the environment.

Climate change concerns human survival. Tackling climate change should thus be an integral part of the government's responsibility to its polity. Global practices demonstrate that governments are the leaders and organizers in the promotion of tackling climate change. Moreover, it is necessary for finance, an essential pillar of state governance, to actively and extensively participate in tackling climate change.

1.2.3 Finance guides expenditure on tackling climate change in various ways

(1) Fiscal policy is an important tool to address climate change

As an important policy tool for the government to carry out economic regulation, fiscal policy has a significant influence on improving market mechanisms and promoting a comprehensive and coordinated development of the social economy. Especially in the context of addressing climate change, since it has the quality of a public good, addressing market failures caused by its externalities must rely on effective fiscal policy measures. Fiscal policy has been a major tool for governments to tackle climate change. It is also important to coordinate and consolidate fiscal, financial, industrial and other policies in developing strategies to mitigate and adapt to climate change.

(2) Fiscal policy is the material guarantee for tackling climate change

Tackling climate change requires huge capital investment. Considering the relatively high risks and uncertainties associated with climate change, the private sector is seldom willing to invest. Therefore, public investment constitutes the main source of funding for tackling climate change. Specifically, taxation is the primary source of government revenue and thus necessary for governments to provide public goods and implement public functions. Apart from stimulating and restricting functions, taxation has the role of raising environmental funds to promote green

---

① Report on Climate Public Expenditure and Institutional Review in China, CAFS, 2015.

growth. The introduction of carbon taxes can provide a stable and adequate source of funding for tackling climate change by improving the environmental resource tax and environmental consumption tax and thus further expanding the environmental tax base. Public financial investment in tackling climate change is also important for stimulating investment and involvement from enterprises and society as a whole. Therefore, fiscal policy offers a guarantee for governments to fund activities to mitigate and adapt to climate change. It is necessary to increase the budget for tackling climate change and gradually establish a stable investment mechanism through which to fund projects.

(3) Fiscal policy has an important role in addressing climate change

The governing role of fiscal policy in climate change adaption mainly concerns its stimulating and restricting functions for energy conservation, energy substitution and new energy development. The stimulating function of fiscal policy in tackling climate change is demonstrated across three aspects. First, it encourages positive externalities, which is to say governments offer corresponding tax relief or financial subsidies to those externalities conducive to resource conservation and environmental protection, and so turn the external benefits into internal economic benefits for the economic entity. Second, through financial subsidies, accelerated depreciation, investment tax compensation and other tax expenditure policies such as subsidies for energy-saving electronic products, increased support for the environmental protection industry and energy conservation sector enables financial capital to guide, encourage and attract social capital investment and establish stable environmental investment channels. Third, it promotes the development and promotion of energy-saving technologies and new environmental protection technologies via financial investment, tax incentives and other measures.

The restricting function is mainly reflected in the restrictions and penalties for waste and on high energy-consuming industries; such restrictions and penalties demand the internalization of negative impacts in order to improve the efficiency of the use of resources and energy, reduce greenhouse gas emissions and achieve sustainable economic and environmental development. Though value-added taxes, resource taxes and consumption taxes in the current taxation structure are not an environmental tax per se, they have played an active role in guiding resource utilization and environmental protection. In the new round of tax reform, consumption taxes will introduce some high pollution and high energy consumption products into the scope of taxation, which will not only help raise funds, but also strengthen environmental protection and resource conservation[①].

1.3 The importance of climate public expenditure and institutional review at the provincial level

1.3.1 To document the effective practices of an ecological civilization in China

The concept of an ecological civilization is unique in China, which when compared with the tradition of using and conquering nature, lays more emphasis on the harmonious coexistence of man and nature. The construction practices of China's ecological civilization are meant to explore a path to modern economic and social development that is in line with traditional Chinese thinking and with the goal to provide support for other countries through Chinese concepts and practices. Local governments are the main implementing agencies in the construction of China's ecological civilization. Climate public expenditure and institutional reviews at the provincial level will help to reproduce the practices of China's ecological civilization construction effectively.

1.3.2 To support decision-making in tackling climate change

Currently, there are relatively few statistical analyses of climate finance in China; an increase in such analysis will assist the government and society to understand the expenditure details of climate finance, improve the budget system and enhance fund efficiency. At the same time, tackling climate change work by local governments is a pilot under the guidance of the central government. More climate public expenditure and institutional reviews of provincial governments could sum up local experience and promote scientific decision-making in tackling climate change.

1.3.3 To compare and exchange view on domestic and foreign climate expenditure

A comprehensive analysis and scientific review of climate public expenditures of local governments in China will help to conduct a comparative analysis of the climate expenditures between China and other countries, evaluate the China's achievements in tackling climate change and in the construction of its ecological civilization, and promote international exchanges.

---

① Report on Climate Public Expenditure and Institutional Review in China, CAFS, 2015.

Report I: Climate Public Expenditure and Institutional Review at the Provincial Level in China——A Study of Hebei Province

2. Climate change and public policies in Hebei

Hebei Province is an area with a dilemma between resources/environmental conditions and development. Those charged with environmental governance are facing enormous challenges. In recent years, however, Hebei has made significant contributions to tackling climate change. To select Hebei as a case study for analyzing and assessing local governments' policies and climate public expenditures can help estimate the overall climate expenditure of local governments in China, provide useful lessons and encourage local governments to tackle climate change effectively.

2.1 The status of climate change in Hebei

2.1.1 The geographic and climatic profile of Hebei

Located in North China, Hebei Province is in the south of the Yanshan Mountains, north of the Yellow River and east of Taihang Mountains. Nestled between the city of Beijing, Tianjin and the Sea of Bohai, Hebei encompasses rich fields to its east, and neighbors Inner Mongolia, Liaoning, Shanxi, Henan and Shandong Provinces. With a coastline 487 kilometers long, it has a total area of 188,800 square kilometers and a total population of 71.85 million. Currently, there are 170 County level districts in Hebei. The province's terrain tilts from northwest to southeast. The northwest parts are mountains, hills and plateaus, among which are basins and valleys. The central and southeastern parts are vast plains. The average altitude of the Bashang plateau falls between 1,200 – 1,500 meters, and the Bashang plateau accounts for 8.5% of the total land area of the province. The Yanshan Mountains and Taihang Mountains, including some hills and basins, are mostly of an elevation below 2,000 meters above sea level, accounting for 48.1% of the total land area of the province. The Hebei plain is part of the North China plain, with an altitude of below 50 meters, accounting for 43.4% of the total land area of the province. Hebei has a temperate continental monsoon climate; there are 2,303.1 hours of the sunshine per annum; the annual frost-free period varies from 81 days to 204 days; the average annual rainfall is 484.5mm; the monthly average temperature is below

3℃ , whereas in July the average is 18℃ to 27℃ ; there are four distinct seasons①.

From the perspective of the trend of climate change, according to the Hebei Implementation Plan for tackling climate change issued by the Hebei provincial government in 2008, the average temperature of Hebei has increased nearly 1.4℃ over the past 50 years; precipitation has dropped by about 120mm; the drought area is expanding at a speed of 1.4% every decade. Such global warming will further intensify in the future. By 2030 the annual average temperature in Hebei will rise by at least 1℃ ; annual precipitation will increase by 3%–13%. By 2050, the yearly average temperature in Hebei will rise by 2.0℃ ; annual precipitation will increase by 3% –15%.

2.1.2 The economy structure and greenhouse gas emissions of Hebei

Hebei is a major industrial province. Its GDP reached RMB2,980.61 billion (US$465.72 billion) in 2015, among which the secondary industry accounted for 48.3%, 7.8 percentage points higher than the national average, and the tertiary industry accounted for 40.2%, 10.3 percentage points lower than the national average.

The main products of Hebei include steel and electro-mechanics. Based on World Steel in Figures 2014 data, world crude steel production in 2013 totalled 1.606 billion tons. As the world's largest steel producer, China accounted for 48.5% of world crude steel production. As China's biggest steel-producing province, Hebei accounted for 11.6% of global crude steel production, far larger than the world's second largest steel producer, Japan, and even higher than the total volume of the 27 member countries of the European Union.

While Hebei is facing enormous challenges in the area of climate change governance, it has also achieved considerable success. According to the China Statistical Yearbook, the total discharge of industrial emissions in Hebei in 2010 was 5,632.4 billion standard cubic meters, accounting for 10.85% of the national emissions in the year, which was more than three times the average discharge of other provinces. The reduction of steel production capacity in 2014 took up 56% of the national task. But by 2015, energy consumption and emissions indicators in Hebei dropped: energy consumption per unit industrial added value was 1.64 tons of standard coal/10,000 yuan, down 33.6% compared with 2010; in 2014, emissions of the main industrial pollutants COD amounted to 151,400 tons, among which amide

---

① About Hebe, data from the official website of Hebei provincial government. http://www.hebei.gov.cn/hebei/10731222/10751792/index.html.

and nitride were 918,000 tons, a 20.7% and 21.8% decrease respectively compared with 2010; the proportion of the main industrial waste gas pollutants (sulfur dioxide, nitrogen oxides, smoke dust) within the total national emissions declined to various degrees; the capacity reduction objective was also successfully completed.

2.2 Planning and regulation

In order to promote tackling climate change in an orderly manner, the Hebei provincial government formulated a series of plans, programs and documents. In 2014, the State Council issued the National Plan on Climate Change (2014 – 2020), which provided measures in nine categories and 40 sub-categories on tackling climate change. We followed this framework to classify the relevant documents issued (abridged) in Hebei (see Table 1 for details).

Table 1 Related documents on tackling climate change issued by Hebei in recent years

	Measures	Related Documents
Control greenhouse gas emission	Adjust industrial structure Optimize energy structure Strengthen energy conservation Increase carbon sinks of forest and ecosystem Control industrial emissions Control emissions from urban and rural construction Control traffic emissions Control emissions in the agriculture, commerce and waste disposal Advocate low-carbon life	The 13th FYP for the Development of Equipment Manufacturing Industry in Hebei The 13th FYP for the Industrial Transformation and Upgrading in Hebei The 13th FYP for the Development of Strategic Emerging Industry in Hebei The 13th FYP for the Development of Petrochemical Industry in Hebei Opinions of the General Office of Hebei Provincial People's Government on the Implementation of Promoting the Development of Biological Industry Comprehensive Implementation Plan of Hebei for Energy Saving and Emission Reduction Opinions of Hebei Provincial People's Government on Implementing the Green Hebei Project Opinions of the Hebei Provincial People's Government on the Implementation of Accelerating the Ecological Restoration of the Forest and Lake The 13th FYP for Meteorological Development in Hebei Opinions of Hebei Provincial People's Government on the Implementation of Technical Transformation of Industrial Enterprises Industrial and Civil Fuel Coal (DB13 / 2081 – 2014) Clean Granular Coal (DB13 / 2122 – 2014) The Outline of the 13th FYP for Housing and Urban-Rural Development in Hebei

Con.

	Measures	Related Documents
Adapt to the impacts of climate change	Improve the resilience of urban and rural infrastructure Strengthen water resources management and facility construction Improve the resilience of agriculture and forestry Improve the resilience of marine and coastal areas Improve the resilience of ecologically fragile areas Improve people's resilience in terms of health Strengthen the system construction of disaster prevention and reduction	The 13th FYP for Water Resource Development in Hebei Methods of Registration of Water Right in Hebei Pilot Project of Comprehensive Control of Groundwater Over-exploitation in Hebei in 2015 Regulations of Hebei on Groundwater Management Outline for the Implementation of Protecting Water Safety in Hebei The 13th FYP for the Development of Modern Agriculture in Hebei The 13th FYP for the Development of Forest Industry in Hebei Opinions of Hebei Provincial People's Government on Implementing the Green Hebei Project The Marine Environmental Protection Plan of Hebei(2016 – 2020) The 13th FYP for the Development of Marine Economy in Hebei The Wetland Conservation Plan of Hebei (2015 – 2020) The 13th FYP for Hygiene and Health in Hebei The Emergency Plan for Heavy Pollution in Hebei Regulations on Prevention and Control of Atmospheric Pollution in Hebei Emergency Plan for Emergency Environmental Events in Hebei
Implement pilot demonstration projects	Deepen the low-carbon pilot projects in provinces and cities Carry out low-carbon park, commercial and community pilot projects Implement carbon reduction demonstration projects Implement pilot projects on tackling climate change	Conducted low-carbon pilot demonstration and circular economy demonstration Construction Plan of New Urbanization and Overall Urban and Rural Demonstration Zone in Hebei (2016 – 2020) Opinions on the Establishment of Clean Production Pilot Demonstration Park Implementation Plan of Air Pollution Control Programme in Hebei Province Regulations on Implementation of Action Plan for Air Pollution Control in Beijing-Tianjin-Hebei and Surrounding Areas

Report I: Climate Public Expenditure and Institutional Review at the Provincial Level in China——A Study of Hebei Province

Con.

	Measures	Related Documents
Improve regional policies on tackling climate change	Urbanized areas Main producing areas of agricultural products Key ecological function areas	The General Plan of High Standard Farmland Construction in Hebei Province (2015 – 2020) Construction Plan of Beijing-Tianjin-Hebei Ecological Environment Support Area in Hebei Province (2016 – 2020)
Improve incentive and restraint mechanism	Promote sound regulations and standards Establish a carbon trading system Establish a carbon emission certification system Improve fiscal and tax policies Improve investment and financing policies	Incentive Measures on Environmental Pollution Reporting in Hebei Province Interim Measures for the Administration of Special Funds for Air Pollution Prevention and Control in Hebei Province (for Trial Implementation) Budget Ration Standard for Restoration and Protection of Geological Environment Project in Hebei Province Measures for the Management of Provincial Environmental Protection Funds in Hebei Province Measures for the Management of Financial Reward Funds on Energy-saving Technological Transformation in Hebei Province Measures for the Management of Special Funds for Provincial Nature Reserves in Hebei Province (for Trial Implementation) Measures for the Management of for Special Funds for Technical Transformation of Industrial Enterprises in Hebei Province
Strengthen the support of science and technology	Enhance the basic research Increase efforts in technology research and development Speed up promotion and application	Measures for the Management of Financial Reward Funds on Energy-saving Technological Transformation in Hebei Province Measures for the Management of Special Funds for Technical Transformation of Industrial Enterprises in Hebei Province
Strengthen capacity building	Improve the greenhouse gas statistical system Strengthen personnel team building Strengthen education and training and public opinion guidance	Carried out statistical work on tackling climate change The 13th FYP for the Development of Human Resources and Social Security in Hebei Province Incentive Measures on Environmental Pollution Reporting in Hebei Province

Con.

Measures		Related Documents
Deepen international exchanges and cooperation	Promote the establishment of a fair and reasonable international climate system Strengthen cooperation with international organizations and developed countries Carry out South-South Cooperation	
Organization and implementation	Strengthen organization and leadership Strengthen co-ordination Establish evaluation mechanism	Implementation Plan for tackling climate change in Hebei Province Opinions of Hebei Provincial People's Government on Accelerating the Construction of Ecological Civilization Implementation Plan for Ecological Civilization System Reform in Hebei Province

Source: Consolidated data from the website of Hebei Provincial Government.

As shown in the table above, Hebei Province has published nearly 50 documents in recent years on tackling climate change. These documents are basically in line with the requirements of the National Plan on Climate Change (2014 – 2020). Except for a lack of corresponding measures concerning international cooperation, relevant plans or regulations have been issued for the remaining areas to guide efforts. Among them, planning documents on the control of greenhouse gas emissions and tackling climate change take up the largest proportion.

In the area of financial expenditure, relevant documents have also been issued. This paper collected 6 documents in total, including the Incentive Measures on Environmental Pollution Reporting in Hebei. The specific content is listed in Table 2.

Table 2　　　　　　　　　　Climate public expenditure in Hebei

Document	Main Content
Incentive Measures on Environmental Pollution Reporting in Hebei	According to the informant's proof on the reported matter, the degree of investigation, the degree of environmental pollution by the party being reported, or the fact that major losses of environmental pollution are avoided because of the reporting, and based on the nature and content of the report, define the amount of reward within the range of 500 – 3000 yuan.

Report I: Climate Public Expenditure and Institutional Review at the Provincial Level in China——A Study of Hebei Province

Con.

Document	Main Content
Interim Measures for the Administration of Special Funds for Air Pollution Prevention and Control in Hebei Province (for Trial Implementation)	The special funds are allocated in a manner that combines projects and factors. Items with specific projects and subsidy standards are assigned in strict accordance with the relevant criteria; other funds are distributed based on factors. The distribution factors mainly take into account the annual reduction of pollutant emissions, investment in pollution control, the reduction rate of fine particulate matter and the assessment results of the previous year, and are defined by integrating with the key tasks in the Implementation Plan of Air Pollution Control Programme in Hebei.
Measures for the Management of Provincial Environmental Protection Funds in Hebei Province	Environmental protection funds provide priority support to the pollution control and ecological protection tasks defined by the nation and Hebei and tend to invest in the ecological environment fields or projects that produce serious environmental pollution and are closely related to people's lives. The allocation of environmental protection funds takes factors as the focus and projects as supplements. Among them, the pre-assigned environmental protection funds based on the amount of tasks will be settled according to the actual completion of the annual targets. The provincial departments applying for the environmental protection funds may, in line with the environmental protection tasks that they undertake, propose fund use plans to the provincial Department of Environmental Protection and prepare performance targets. The provincial Department of Finance and Department of Environmental Protection will consider the relevant environmental tasks and coordinate arrangements for environmental protection funds. The municipal bureaus of finance and environmental protection shall, after receiving the environmental protection funds, promptly prepare a detailed allocation plan for the environmental protection funds, and shall, upon approval according to the prescribed procedures, issue it within 30 days and submit it to the provincial Department of Finance and Department of Environmental Protection for the record. The municipal bureaus of finance and environmental protection and relevant provincial departments that use the environmental protection funds shall conduct a performance assessment of environmental protection funds based on projects and form a special evaluation report to be submitted to the provincial Department of Finance and the Department of Environmental Protection for the record. Based on the performance assessment of the local agencies, the provincial Department of Finance and the Department of Environmental Protection conduct key performance assessment.

Con.

Document	Main Content
Measures for the Management of Financial Reward Funds on Energy-saving Technological Transformation in Hebei	In order to ensure the actual effect of energy-saving technological transformation projects, energy-saving funds take the form of incentives, linking the amount of funds with the volume of energy saving by the projects and providing rewards to enterprises that implement energy-saving technological transformation projects. The incentive funds focus on the key energy-consuming enterprises implementing energy-saving technological transformation projects for existing production technology and equipment; incentive funds mainly support the enterprises' energy-saving technological transformation projects and the amount of incentives is based on the actual energy saving; the energy conservation is reported by the company and confirmed by the government. Enterprises submit the energy consumption before the transformation, energy-saving measures after the conversion, energy conservation and measurement methods, which will be identified by the government based on professional audit results. The performance indicators of the funds are set by the provincial Development and Reform Commission in line with the overall objectives of a project to determine the project's target values and report to the provincial Department of Finance for audit. The performance indicators include fund management indicators, output indicators and effect indicators. Specifically, fund management indicators include capital management norms and fund availability; output indicators include the number of project implemented, investment scale and energy efficiency; effect indicators include economic and social benefits.
Measures for the Management of Special Funds for Provincial Nature Reserves in Hebei (for Trial Implementation)	The special funds are applied for: (1) comprehensive field investigation and the planning formulation for the development and construction of reserve areas; (2) procurement of equipment for scientific research, observation, monitoring and early warning in nature reserves; (3) facilities as well as scientific researches and tests for rare and endangered species and biodiversity conservation; (4) publicity of natural ecological protection; (5) other matters conducive to the development of nature reserves. The special funds for nature reserves should set the performance indicators based on the performance objectives, including reporting indicators, fund management indicators, output indicators and effect indicators. Among them, reporting indicators include the integrity of application materials as well as the independence and soundness of management institutions in the reserve area; fund management indicators include normative management of the special funds and fund availability; output indicators include the number and type of management facilities and equipment, the number and quality of planning, scientific research experiments, and publicity materials; effect indicators include the operation effect of the project, protection of wild animals and plants, and protection of ecological environment.

Report I: Climate Public Expenditure and Institutional Review at the Provincial Level in China——A Study of Hebei Province

Con.

Document	Main Content
Measures for the Management of Technical Transformation for Industrial Enterprises in Hebei	The special funds mainly support the annual Transformation of Thousand Technologies Project, focusing on 10 - 100 - 1000 Project①, 3 - 100s Project, Model Project and the model enterprises. The special funds provide comprehensive forms of support, including post subsidies, loan discount, equity investment and service procurement. Post subsidies refer to a certain amount of subsidies provided to the major technical transformation projects as leading role models in the industrial transformation and upgrading in Hebei-based on their investment in procuring fixed assets including equipment, and the maximum subsidy for a single project is 20 million yuan; investment in the fixed assets of public platform projects should not exceed 20% of the subsidy, and for a single project it should not exceed 5 million yuan. The discount rate shall be calculated based on the actual amount of the project loan and the benchmark interest rate for loan of the same period announced by the People's Bank. The amount of support shall not exceed 12 months' interest and the maximum amount for a single project shall not exceed 20 million yuan.

Source: Consolidated data from the website of Hebei Provincial Government.

For policies related to climate public expenditure in Hebei, see ANNEX I for detailed information. The information was gathered during field investigations and desk reviews of secondary sources made available to the public.

2.3 Climate change institutional set-up in Hebei

The Hebei provincial party committee and provincial government attach great importance to climate change and desire to set up a leading task force to tackle climate change in February 2008 with the governor as team leader. A response mechanism was established, and the Implementation Plan for tackling climate change in Hebei was formulated in accordance with the requirements of sustainable development strategy. A series of policies and measures related to climate change have been

---

① The Transformation of Thousand Technologies Project is a key technological transformation program for provincial industrial enterprises. It was established by Hebei Province in 2014 and is updated per annum. These projects have started or have basic start conditions. The project's implementation will be of great significance for Hebei to adjust their industrial structure, change their mode of development and stabilize economic growth. The subsequent 10 - 100 - 1000 Project are the implementing measures put forward by the provincial government to promote industrial transformation and upgrading and to build a strong industrial province, which includes the strengthening of 10 major industrial bases, supporting of 100 advanced enterprises and the cultivation of 1000 brand-name products. The 3 - 100s Project refers to the 100 continued construction and operation projects, 100 new start projects and 100 pre-projects, through which Hebei province aims to expand effective investment.

adopted, and positive efforts have been made to mitigate and adapt to climate change. The Hebei Provincial Development and Reform Commission has set up a division of tackling climate change, which is responsible for the day-to-day work of the Office for the Provincial Leading Task Force on Tackling Climate Change (see ANNEX II for a detailed description of terms and responsibilities of the varied government departments).

Figure 1　The macro coordination mechanism for tackling climate change in Hebei

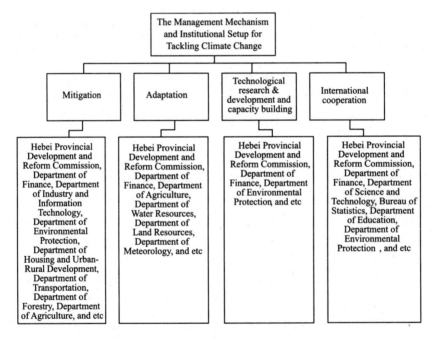

Figure 2　The management mechanism and institutional setup for tackling climate change in Hebei

3. Analysis of the budget system of climate public expenditure in Hebei

3.1 The structure of budget management

China has varied levels of governance and at each level functions are performed with sufficient financial support. Therefore, each level of governance has a main budget management body which takes charge of fiscal revenue and expenditure. Article 3 of the Budget Law of the People's Republic of China stipulates that the state shall establish one budget level at each of the five levels of governance: central government; provinces, autonomous regions, and municipalities directly under the central government; cities divided into districts and autonomous prefectures; counties, autonomous counties, cities not divided into districts, and districts of a city; townships, ethnic townships, and towns. Hebei has four budget levels: provinces, cities, counties and townships.

Table 3   Government budget management bodies

Level	Budget Management Government
I	Central government
II	Provinces, autonomous regions, and municipalities directly under the central government
III	Cities with districts and autonomous prefectures
IV	Counties, autonomous region, cities without divided districts, and districts of a city
V	Townships, ethnic townships, and towns

3.2 Budget preparation

Departmental budgets are the basis for the general budget preparation of governments at all levels. According to the Budget Law of the People's Republic of China, government budget preparation follows a combination of top-down, bottom-up, and level-by-level procedures.

Provinces, autonomous regions and municipalities directly under the central government propose the requirements for preparing a draft budget in their administrative area; such proposals must be in line with the instructions of the State Council and the Ministry of Finance, and take into account the specific circumstances of the region.

For a local government at or above the county level, its financial department shall review the draft budgets of all the departments and prepare the draft budget and general budget at the same level. After the approval of the government of the same level, they will prepare regular reports for the financial departments of the provinces, autonomous regions and municipalities directly under the central government at the next level up. The draft general budget prepared by the financial department of the same level should be reported to the Ministry of Finance before January 10 of the next year.

The draft budgets of local governments at all levels shall be approved by the people's congress of the same level. The financial departments of local governments at or above the county level shall, within 30 days from the date of government budget approval by the people's congresses at the same level, approve the budgets of the departments at the same level. Local departments shall, within 15 days from the date of budget approval by the financial department at the same level, approve the budgets of their respective units.

3.3 Budget item setting

In order to strengthen revenue management and statistical analysis, according to the revenue composition of the Chinese government, and with reference to international classification methods, financial budget revenue is divided into class, section, item and head, based on their economic nature; the expenditure is divided into class[1], section[2] and item[3] according to functions, and class and section according to their economic nature. Budget expenditures at all levels should be compiled according to their functional and economic natures.

3.3.1 Revenue classification

According to the requirements concerning government revenue, government

---

[1] It comprehensively reflects governmental functions such as defense, diplomacy, education, science and technology, social security and employment, environmental protection and so on.

[2] It reflects a certain aspect of the work needed to complete governmental functions, such as general education under education.

[3] It reflects the specific expenditure spent to complete a particular aspect of work, such as drought prevention and soil and water conservation under water conservancy expenditure.

Report I: Climate Public Expenditure and Institutional Review at the Provincial Level in China——A Study of Hebei Province

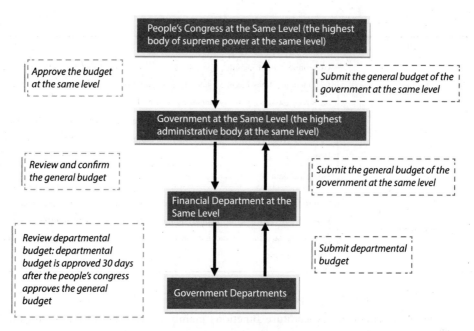

Figure 3 Budget preparation flow chart

revenue is divided into class, section, item and head, all of which are refined gradually, item by item, to meet the different levels of management needs. Government revenue is divided into tax revenue, revenue from social insurance funds, non-tax revenue, principle recovery income from loan re-lending, debt revenue and transfer income. Tax revenue is divided into 21 sections, revenue from social insurance funds into 6 sections, non-tax revenue into 8 sections, principle recovery income from loan re-lending into 4 sections, debt revenue into 2 sections, and transfer income into 10 sections.

Table 4 Budget revenue item settings

Class	Section
Tax Revenue	Value-added tax, consumption tax, business tax, corporate income tax, returned tax of corporate income tax, individual income tax, resources tax, fixed assets investment orientation regulation tax, urban maintenance and construction tax, property tax, stamp tax, urban land use tax, land value increment tax, vehicle use and license tax, tonnage tax, vehicle purchase tax, tariff, farmland occupation tax, deed tax, other tax revenues
Revenue of social insurance funds	Basic pension insurance fund income, unemployment insurance fund income, basic medical insurance fund income, work injury insurance fund income, maternity insurance fund income, other social insurance fund income

Con.

Class	Section
Non-tax revenue	Government fund income, special income, lottery fund income, administrative fees income, fines income, state-owned capital operating income, state-owned resources (assets) paid use income and other income
Principle recovery for loan re-lending	Domestic principal recovery for loans, foreign principal recovery for loans, domestic principle recovery for loan re-lending, foreign principle recovery for loan re-lending
Debt revenue	Domestic debt revenue and foreign debt revenue
Transfer income	Return income, financial transfer income, special transfer payment income, government fund transfer income, lottery public welfare fund transfer income, extra-budgetary transfer income, income balance from the previous year, and transferred fund

3.3.2 Classification of expenditure functions

The classification of expenditure functions mainly reflects the different roles and policy objectives of the government's functional activities, and explains what the government has done. Government expenditure functions are divided into class, section and item. Class reflects the comprehensive functions of the government; section reflects the work needed to complete government functions; item refers to the specific expenditure spent to complete a particular aspect of the work. Items of these three levels are set up in a sequence from large to small, from coarse to fine, and in a hierarchical way. Based on functions, general public budget expenditure is divided into 17 classes, over 170 sections and more than 800 items, which include expenditure on general public service, foreign affairs, public safety, national defense, agriculture, environmental protection, education, science and technology, culture, health, sports, social security, employment and others.

4. Statistics and assessment of climate public expenditure in Hebei

4.1 The determination of the scope of climate public expenditure in Hebei

Climate public spending is broadly categorized into three areas for further analysis as indicated below:

4.1.1 Expenditure on climate change mitigation

- Energy restructuring: support to renewable energy, clean coal technology and nuclear power.

- Industrial restructuring: elimination of backward steel production capacity; development of special funds for strategic emerging industries; support for technological upgrading and structural adjustment projects for traditional industries.

- Energy saving and efficiency improvement: expenditures to promote energy saving, the comprehensive utilization of resources and ecological construction.

- Development of emissions trading: the technical and transactional work of the major pollutant emissions trading in the province, the construction, management and maintenance of the province's emissions trading network and platform and related services provided to emissions trading activities.

4.1.2 Expenditure on tackling climate change

- Agriculture: expenditures on the protection, restoration and utilization of agricultural resources, disaster prevention and reduction and grassland vegetation restoration.

- Water conservancy: expenditures on the management and protection of water resource conservation and and water conservancy construction.

- Meteorology: expenditures on climate change monitoring, forecasting and assessment.

- Urban infrastructure construction: expenditures on urban sewage treatment facilities and the construction of supporting pipe networks, the sustainable treatment of urban garbage and the construction of supporting facilities, energy-saving building materials and the greening of urban planning areas.

4.1.3 Expenditure on climate capacity building

- Institution and personnel: operation and expenditure of public climate sectors and agencies; expenditure on the establishment of an expert pool for the environmental protection fund in Hebei.

- Research and development expenditure.

- Expenditure on environmental publicity and education.

4.2 Statistics on climate public expenditure in Hebei

4.2.1 Basic research ideas

To acquire information on the specific practices used in producing climate statistics, this report plans to refer to the approach adopted by the first phase of CPEIR. Based on the actual situation in Hebei, this report then plans to use government revenue and expenditure classifications as our primary research method, while simultaneously screening for specific expenditure items. To begin, this report intends to search for climate-related budget items according to government revenue and expenditure classifications, and then make detailed classifications by combining the expenditure items under various subjects and referring to climate correlation.

4.2.2 Classification of provincial public expenditure on its climate relevance

Based on classification methodology provided by the United Nations Development Program (UNDP) and adopted in many other developing countries in the Asia-Pacific region, and within the context of the current revenue and expenditure classification in Hebei (class, section and item), this report divided the fiscal expenditures based on climate correlation.

(1) High correlation: the primary objective of expenditure is directly related to tackling climate change, including environmental monitoring and supervision, pollution prevention and control, natural ecological protection, natural forest protection, returning farmland to forests, sandstorm management, returning grazing land to grassland, energy conservation, pollution discharge reduction, renewable energy and the comprehensive utilization of resources under energy saving and environmental protection; natural disaster relief under social security; disease prevention and control, response to public health emergencies, basic public health services, and major public health special projects under public health; disaster relief and protection and utilization of agricultural resources under agriculture; forestry; disaster prevention and reduction and other expenditures.

(2) Moderate correlation: the secondary targets of expenditures are related to tackling climate change, including expenditures on water conservancy and the South-to-North Water Diversion under agriculture, forestry and water resources; energy management affairs under energy conservation and environmental protection; development and reform affairs, and environmental protection and management affairs under general public services; basic research under science and technology expenditure; expenditures for emergency treatment agencies, other professional public health institutions and maternal and child health care institutions under public health; expenditures on the planning and protection of key national scenic areas, planning and management of urban and rural communities, facilities of urban and rural communities under urban and rural communities; railway expenditure under transportation and other

Report I: Climate Public Expenditure and Institutional Review at the Provincial Level in China——A Study of Hebei Province

expenditures.

(3) Low correlation: includes or involves expenditures related to tackling climate change, such as financial affairs and statistical information affairs under general public services; science and technology management affairs, applied research, scientific and technical conditions and services, and the popularization of science and technology under science and technology expenditure; urban and rural communities management under urban and rural communities expenditure; expenditures on agricultural administrative operation and poverty alleviation in agriculture, forestry and water resources; highway and waterway transportation under transportation; public security organs under public security expenditure.

(4) No correlation: includes expenditures which have no direct connection with tackling climate change, such as the expenditures on the people's congress affairs, People's Political Consultative Conference affairs, human resources affairs and national defense under the general public services. Here, no correlation also includes unclassifiable expenses such as spending on education, culture, sports and media.

Table 5 Climate relevance of provincial general public budget expenditure (abridged)

NO.	Item	High Correlation	Moderate Correlation	Low Correlation	No Correlation
1	General public services expenditure		√	√	√
2	National defense expenditure		√	√	√
3	Public security expenditure				√
4	Education expenditure				√
5	Science and technology expenditure		√	√	
6	Culture, sports and media expenditure				√
7	Social security and employment expenditure		√		√
8	Health care expenditure	√	√	√	√
9	Energy saving and environmental protection expenditure	√	√		
10	Urban and rural communities expenditure			√	√
11	Agriculture, forestry and water resources expenditure	√	√	√	
12	Transportation expenditure				√
13	Resource exploration information expenditure			√	√
14	Commercial and services expenditure				√
15	Fiscal expenditure				√
16	Aid to other areas				
17	Land resources, marine and meteorology and other expenditure	√	√	√	

Con.

NO.	Item	High Correlation	Moderate Correlation	Low Correlation	No Correlation
18	Housing security expenditure				√
19	Cereal and oil reserve expenditure				√
20	Preparation fees				√
21	Other expenditures				√

According to the abridged table above, energy saving and environmental protection expenditures and agriculture, forestry and water resources expenditures have high correlation with climate change. Taking these two as cases, this report conducted an analysis on government spending at different levels.

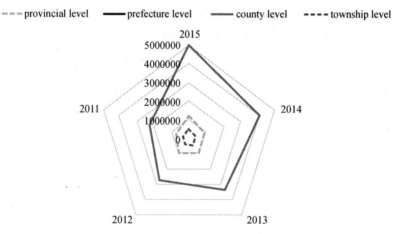

Figure 4 Expenditure on energy saving and environmental protection at all government levels in Hebei from 2011 – 2015 (10,000 RMB)

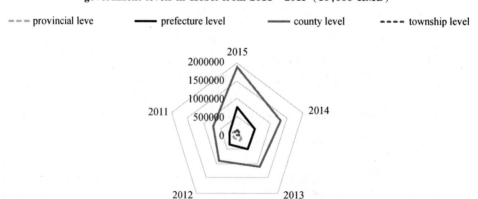

Figure 5 Expenditure on agriculture, forestry and water resources at all government levels in Hebei from 2011 – 2015 (10,000 RMB)

Report I: Climate Public Expenditure and Institutional Review at the Provincial Level in China——A Study of Hebei Province

Figure 4 and Figure 5 show that in the expenditures on energy saving and environmental protection and agriculture, forestry and water resources, county governments have invested the most, followed by municipal government. Village and town governments, as well as provincial governments, have invested proportionately less.

This report has classified the general public budget expenditure of Hebei, according to its climate relevance (refer to ANNEX IV).

4.2.3 Statistical results and analysis

According to the correlation analysis of climate expenditures, the proportion of expenditure with high climate correlation in Hebei in total public spending of Hebei between 2001 and 2015 was 4.78%, 4.69%, 5.49%, 6.02% and 6.93%, respectively. The proportion of climate expenditures with moderate climate correlation in the provincial fiscal expenditure was 4.34%, 4.06%, 4.02%, 4.44% and 4.35%, respectively. The proportion of climate expenditures with both high climate and moderate correlation was 8.23%, 8.75%, 9.51%, 10.46% and 11.29%, respectively. See Figure 5 for details. The data shows that Hebei attaches great importance to climate change. During the 12th FYP period, climate public expenditure increased year over year. From the perspective of the proportion of climate public expenditure in total provincial fiscal expenditure, the proportion is higher than the central climate-related expenditure of the same range. According to the statistical analysis of the first phase, the proportion of climate expenditures with high and moderate correlation at the central level was 7.6% and 6.9% in 2013 – 2014[1].

Fiscal spending, which is highly relevant to climate, is on the rise; expenditures of moderate climate correlation rose from 3.44% in 2011 to 4.35% in 2015; expenditures of low climate correlation are declining (see Figure 6 for details).

According to the above-mentioned proportion of climate-related expenditures, the size of expenditures on an annual basis can be obtained. Based on the calculation, expenditures with high climate correlation in Hebei between 2011 and 2015 were RMB16.92 billion (US$2.64 billion), RMB19.145 billion (US$5.64 billion), RMB24.225 billion (US$3.79 billion), RMB28.155 billion (US$4.40 billion), and RMB39.05 billion (US$6.10 billion), respectively. Data shows that during the 12th FYP period, climate-related expenditures in Hebei increased year over year. Expenditures with high climate correlation in 2015 increased by 38.7% over the

---

[1] Report on Climate Public Expenditure and Institutional Review in China, CAFS, 2015.

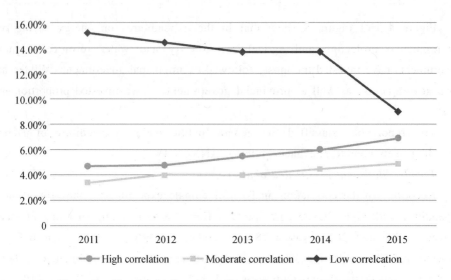

Figure 6　Changes in the proportion of fiscal expenditure in Hebei by climate correlation

previous year, mainly due to greater investment in energy saving, environmental protection and other measures (see Figure 7 and Figure 8 for details).

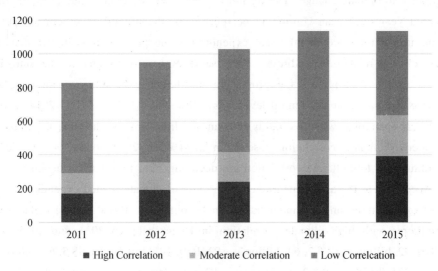

Figure 7　Climate public expenditure in Hebei between 2011 – 2015

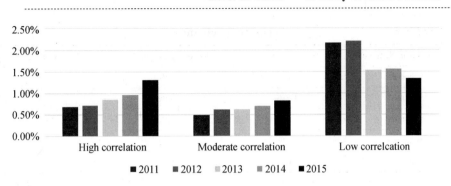

Figure 8　Proportion of climate-related expenditure in GDP of the corresponding year in Hebei between 2011 – 2015

As shown in Figure 7, among the general public expenditure in Hebei, the size of climate-related fiscal expenditures is on the rise. As shown in Figure 8, the proportion of expenditure with high or moderate climate correlation to the GDP of the corresponding year in Hebei demonstrates an upward trend, while expenditures with low climate correlation are following a downward trend. The primary reason for this phenomenon is that Hebei has made reasonable adjustments to their climate-related fiscal expenditures, investing more fiscal funds in areas with high correlation to climate change, and enhancing the pertinence and efficiency of fund use.

According to the degree of climate correlation, and combined with the specific fiscal expenditure situation in Hebei, it is important to weight the high, low and moderate correlations, and conduct more accurate analysis on climate related expenditures (Table 6). Based on each climate correlation's respective assignment scheme, the total proportion of climate public expenditure in fiscal expenditure in Hebei is calculated (Table 7). The results show that in 2015, the proportion of climate public expenditure within the total fiscal expenditure in Hebei is 12.09% at the most optimistic estimation, and 8.44% at the most conservative estimation.

4.2.4　*Further reflection on existing analytical methods*

This report believes that there is room for further improvement in the existing analytical methods, and that the criteria of correlation assessment needs to be further adjusted. For example, when this report defines the climate correlation of an activity by the criteria of climate change mitigation and climate change adaption, the results will be different; such results will affect the statistics of relevant fiscal funds.

The two main aspects of addressing climate change are mitigation (reducing greenhouse gas emissions) and adaptation (recognizing climate change and establishing mechanisms to strengthen resilience). Both developed and developing

Table 6   Weighting schemes①

Assignment Scheme	High Correlation	Moderate Correlation	Low Correlation
Scheme 1	100%	70%	30%
Scheme 2	100%	50%	20%
Scheme 3	80%	50%	20%
Scheme 4	70%	50%	20%
Scheme 5	90%	50%	20%

Table 7   Analysis of the proportion of climate-related expenditure in fiscal expenditure in different schemes

Assignment Scheme	2011	2012	2013	2014	2015
Scheme 1	11.69%	11.85%	11.26%	12.08%	12.09%
Scheme 2	9.50%	9.60%	9.47%	10.21%	10.52%
Scheme 3	8.55%	8.66%	8.37%	9.01%	9.13%
Scheme 4	8.07%	8.20%	7.82%	8.40%	8.44%
Scheme 5	9.03%	9.13%	8.92%	9.61%	9.82%

countries must adopt "measurable, reportable and verifiable" mitigation actions. Mitigation measures are aimed at addressing the causes of climate change, while adaption focuses on addressing the impact of climate change. Adaptation refers to the adoption of policies and practices to address the impacts of climate change. Different sectors have different adaptation measures, such as increasing water resources by expanding rainwater collection, water storage and water conservation, bolstering agricultural production by adjusting planting dates, crop varieties and engaging in crop relocation, increasing infrastructure construction (including in coastal areas) by establishing wetlands as barriers against sea level rise and flooding, optimizing energy utilization by using renewable energy and improving energy efficiency.

According to the definition of climate change mitigation and adaptation, combined with the practices of Hebei in China, this report re-assessed the climate correlations of the relevant budget expenditure items (see Annex V, table 1, 2 and 3

---

① The above percentage has a certain degree of simulation. As it is impossible for further breakdown and statistical analysis on climate public expenditure from open source, based on the relevant national statistical experience and the differences among expenditures in terms of climate correlation, make predictions on the proportion of climate expenditure. As shown in Scheme 1, it is assumed that for the fiscal expenditure with high climate correlation, 100% of the fiscal expenditure is used to cope with climate change, and 70% for expenditures with moderate climate correlation and 30% for expenditures with low climate correlation. So are other schemes.

Report I: Climate Public Expenditure and Institutional Review at the Provincial Level in China——A Study of Hebei Province

for details).

The re-assessment results show: in the case of high correlation, the number of climate change mitigation items accounts for 36% of the total number of items, the number of tackling climate change items accounts for 25%, while the number of items with both correlations accounts for 39%; in terms of moderate correlation, the number of climate change mitigation items accounts for 19% of the total number of items, the number of tackling climate change items accounts for 42%, while the number of items with both correlations accounts for 38%; in the area of low correlation, the number of climate change mitigation items accounts for 20% of the total number of items, the number of tackling climate change items accounts for 36%, while the number of items with both correlations accounts for 44% (see Figure 9 for details).

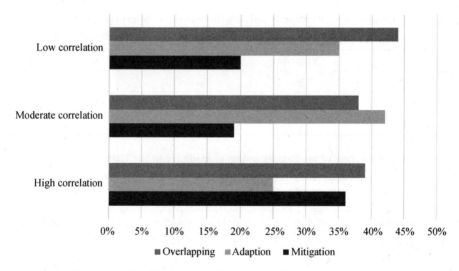

Figure 9   Re-classification results of general public budget expenditure items according to climate correlation

Taking the 2015 general public expenditure in Hebei as an example, based on the re-classified climate correlation standards, re-calculation was conducted. The results show that in the case of high correlation, the overlapping amount of mitigation and adaptation accounts for 21.19% of the expenditure; in the case of moderate correlation, the overlapping amount of mitigation and adaptation accounts for 20.02% of the expenditure; in the case of low correlation, the overlapping amount of mitigation and adaptation accounts for 56.06% of the expenditure (see Table 8 for details). Analysis of the specific items indicates that the overlapping parts are mostly a result of institutional operating expenditures, research and development expenditures and publicity spending. These activities share the functions of mitigation and adaptation. Refer to Table 9 and 10 for the proportion of the climate public expenditure, according to the reclassified statistics, in the general public expenditure and GDP of the corresponding year.

Table 8　　Statistics of climate-related expenditure in Hebei in 2015 after reclassification　　Unit: Billion Yuan

	Mitigation	Adaptation	Mitigation & Adaptation	Percentage of Overlapping (%)
High Correlation	33.661	13.667	8.273	21.19
Moderate Correlation	20.95	8.485	4.911	20.02
Low Correlation	44.524	33.52	28.044	56.06

Table 9　　Proportion of the climate public expenditure in the general public expenditure of the corresponding year in Hebei after the reclassification

	Mitigation	Adaptation	Mitigation & Adaptation
High Correlation	5.98%	2.43%	1.47%
Moderate Correlation	3.72%	1.51%	0.87%
Low Correlation	7.91%	5.95%	4.98%

Table 10　　Proportion of the climate public expenditure in the GDP of the corresponding year in Hebei after the reclassification

	Mitigation	Adaptation	Mitigation & Adaptation
High Correlation	1.13%	0.46%	0.28%
Moderate Correlation	0.70%	0.28%	0.16%
Low Correlation	1.49%	1.12%	0.94%

The above statistical analysis shows that more than half of the expenditure items with high, moderate and low correlation can be subdivided again based on the definition of climate change mitigation and adaptation. After the re-classification, the amount of fiscal expenditure used to tackle climate change can be calculated more accurately.

5. Key Conclusions and Recommendations

5.1 Key Conclusions

5.1.1 With increasing importance being attached to climate change, climate public expenditures have grown year over year.

Since the 18th National Congress of the CPC, China has always seen the development of an ecological civilization as an important strategy for the governance of the country. The Third Plenum of the 18th National Congress of the CPC proposed to accelerate the development of systematic and complete ecological civilization systems; the Fourth Plenum required that stringent legal systems be adopted to protect the ecological environment; and the Fifth Plenum introduced green development as a new development concept. Hebei has also attached great importance to climate change. A review of the province's climate public expenditures shows that its expenditures have been growing year over year throughout the 12th FYP period. After 2015, with increased investment in energy conservation and environmental protection, Hebei's public expenditures highly relevant to climate change grew by 38.7% over the previous year. Public expenditures highly or moderately relevant to climate change has been on the rise during the period of 2011 – 2015, with the share of such expenditure within the province's total expenditure having increased from 8.23% to 11.29%. This proportion is even higher than that of the central level within a the same statistical scope.

5.1.2 Climate change finance is characterized by multiple sources of funding, diversified methods of investment and extensive areas of investment.

The CPEIR of Hebei indicates that the province's financial contribution to addressing climate change is derived from many sources, including the government, capital markets and international loans. Apart from the general public expenditure, fiscal input from the government also includes state-owned capital and government-owned expenditure. The government has explored the use of market-oriented instruments (e.g., emissions trading) and financial instruments (e.g., green securities) to support emerging industries, environmentally-friendly industries and low-carbon industries; they have also explored the provision of special funds and the replacement of subsidies with incentives. Fiscal investment covers both climate change mitigation and adaptation. The measures to mitigate climate change have shifted from ex-post coping to prevention and control of effluents at the source. Adaption measures cover the setup and functional adjustment of administrative organizations, and advocacy for and changes in human health, lifestyle, travel means and production modes.

5.1.3 The improvement of climate public expenditure budget system provides institutional guarantees for enhancing the efficiency of fund uses and for strengthening fund supervision.

The budget of a government reflects the scope of the government's activities, as well as the policies to be achieved by the government over a particular period; the budget system is a safeguard for the government to attain its policy goals. As the reform of budget management systems deepens, the budget items for climate change public expenditure have been improving. In 2015, for example, "energy conservation and environment protection" was added to the government's list of budget expenditure items, and items highly relevant to measures that reflect and address climate change were added to the fiscal budget; a full coverage budget management system has also been preliminarily established to serve as a coordination and integration mechanism for the budget of government-managed funds, the budget of state-owned capital operations, the budget of social insurance funds and that of the general public budget. Moreover, efforts have been exerted to enhance the coordination of funds from governments at or above the local level, and to incorporate off-budget funds into the budget to improve the transparency of fund uses and facilitate better supervision. Budget control has also been improved. For example, medium-term fiscal plan management and the intertemporal budget balance mechanism have enhanced the integration of budgeting and planning, thus improving the appropriation and utilization of fiscal funds, reducing the size of idle fiscal funds and contributing to the efficiency of fund uses.

5.1.4 The statistical and evaluation methods for climate public expenditure have been expanded.

In the first stage of CPEIR, public spending was statistically analyzed by category in light of the relevance of such funds to climate change activities; this was done in order to evaluate the scale of climate change public expenditures at the central government level. In the second stage, however, this report has expanded the categorization and statistical analysis methods for climate public expenditure. This report further broke down provincial expenditures into those for climate change mitigation and those for adaption. It found that this subdivision-based statistical analysis method gathered more precise data on the expenditures directed to address climate change, thus laying a foundation for an accurate cost-benefit analysis and performance appraisal of public expenditures in general.

5.2 Recommendations

5.2.1 To address climate change requires an enhancement of the dissemination of concepts and top-level policy designs to foster virtuous interactions between the government and the market.

At the national level of governance, development concepts such as the "lucid waters and lush mountains are invaluable assets"[①], "we need not only lucid waters and lush mountains, but also gold and silver mines" and "we'd rather have lucid waters and lush mountains than have gold and silver mines" have been reiterated again and again. At the operational level, despite decrease in tax revenue and economic downturns due to the lack of alternative drivers of growth, Hebei has exerted certain supervision of negative measures that fuel climate change. Moreover, they resort to administrative means to do so, while respecting market principles. In policy design, therefore, there is a need to clarify concepts, heed the development of top-level institutions and use systematic thinking to guide climate change activities. Governments must use market-oriented means more consistently to address climate change. Moreover, measures such as raising the entry threshold for industries, imposing taxes and levying charges can be adopted to internalize the cost of pollution of enterprises and increase their cost, thereby removing their incentive to continue the exploitive development path. In addition, by using the laws of pricing, governments can increase their purchase of energy-saving and consumption – reducing products to impact market demand and in turn promote the transformation and upgrading of production.

5.2.2 There is a need to deepen the reform in budget management systems in order to improve the effectiveness of policies and funds.

To address climate change, Hebei has increased public spending and made many efforts to explore ways to increase its effectiveness. It is suggested that budget management system is further improved. On the one hand, fiscal budget management

---

① http://www.cctb.net/bygz/zywxsy/201511/t20151113_331161.htm.

must introduce the "medium – term" concept and practice medium-term fiscal plan management in order to make the budgetary arrangement align with the medium-to-long-term policies of the government, shift governmental decisions on climate change activities from annual ones to more forward – looking and continuous medium-term ones, and ensure that current policies support the long – term sustainability of public finance. On the other hand, there is the need to practice whole process budget performance management, improve the quality of performance information and integrate performance management with appraisal results and budgetary arrangement. In creating budgets, there is the need to set performance goals and establish standardized performance indicator systems. Quantitative performance indicators should include not only such indicators as the quantity, quality, timing and cost of output, but also such indicators as the economic, social and ecological benefits, the impact on sustainability and the satisfaction of service recipients. Finally, there is a need to conduct performance appraisals on climate public expenditure policies, prepare performance appraisal reports for such policies and disclose them to the general public.

5.2.3 The methods of fiscal expenditure need to be further innovated to establish market-oriented restraint mechanisms.

To address climate change is an arduous task. The result indicates that the climate related public spending has been growing in Hebei in recent years. Meanwhile, greater attention has been given to reform innovative spending modalities and improve cost-effectiveness. However, in contrast to the enormous financial demand, fiscal funds may be far from enough to tackle climate change. In addition to widening channels of financial input, there is also the need to innovate, to enhance the integration of public finance with the use of financial instruments (e. g. , funds, bonds), and to enhance the leverage capacity of fiscal funds. For example, the government may attract social capital into climate change activities by way of PPP models, fiscally subsidized interest rates and green finance. Moreover, the government may use market – oriented means (e. g. , emissions trading) and financial instruments (e. g. , green bonds) to support the development of emerging, environment-friendly and low carbon industries.

5.2.4 Research on the cost-benefit analysis for climate public expenditure needs to be enhanced.

The review of climate public expenditure is a new area of research. In addition to establishing scientific categorization and statistical analysis methods to evaluate the

Report I: Climate Public Expenditure and Institutional Review at the Provincial Level in China——A Study of Hebei Province

total size and structure of climate public expenditure, there is also the need to conduct comprehensive cost-benefit analysis for such expenditure; such analysis is to identify the cost to all stakeholders involved in tackling climate change and the eco-environmental, social and economic benefits of climate public expenditure, thereby providing a basis for improving climate public expenditure policies.

Report II:
Cost-Benefit Analysis of Climate Public Expenditure
——A Case Study on Overcapacity Reduction in Hebei Province

Introduction

Climate change is closely linked to a country's mode of economic development. Facilitating the transformation of an economic development mode towards low-carbon development mode is necessary to deal with climate change. At present, China is pushing forward supply-side structural reform, overcapacity reduction being an important part of such a reform. Existing overcapacity, which is a result of extensive development in the past, has not only increased the pressures of the present economic downturn, but also resulted in a number of defects in environmental protection that could bring an end to China's sustainability efforts. As an important measure for dealing with climate change, overcapacity reduction would optimize the production structure, improve the efficiency of resource utilization, reduce industrial emissions and improve the quality of economic development.

Hebei Province has been typical in its response to climate change. As a major industrial province, its GDP in 2016 reached RMB3.18 trillion (US$ 497.31 billion), with the added value of secondary industries being RMB1.51 trillion (US$ 235.24 billion). These numbers accounted for 47.3% of the GDP and were 7.5%

Report II: Cost-Benefit Analysis of Climate Public Expenditure ——A Case Study on Overcapacity Reduction in Hebei Province

higher than the national average level. Industry occupies an important position in the economy of Hebei. Hebei is a big producer of iron and steel. In 2016, the output of pig iron was 183.94 million tons, the output of crude steel was 192.60 million tons, and the output of steel products was 261.504 million tons. The output of iron and steel accounted for 1/4 of the total output of the whole country. In recent years, Hebei has actively responded to climate change, consistently increasing climate-related public expenditures, and notable results have been achieved.

Based on the CPEIR in Hebei, this report includes cost-benefit analysis on specific projects of public expenditure concerning climate change. The aim is to explore the methodological framework of climate public expenditure performance review. In view of the fact that overcapacity reduction has been one of the important measures to deal with climate change in Hebei over recent years, this report has selected the project of overcapacity reduction in Hebei to conduct the research. Taking into account our field visits and investigations of relevant governmental departments and enterprises in Hebei, this report has conducted a relatively systematic analysis of the costs and benefits of overcapacity reduction, exploring the construction of a methodological framework for cost-benefit analysis on climate public expenditure.

1. Progress made in overcapacity reduction in Hebei

In September 2013, President Xi Jinping delivered a speech on resolving excess capacity, and commissioned Hebei to be leader in this effort. In 2013, Hebei formulated and implemented its *Implementation Plan on Resolving Serious Production Overcapacity Conflicts*, which put forward the "6643 Program" to resolve overcapacity. By 2017, according to the "6643 Program", the production capacity related to the production of 60 million tons of steel, 60 million tons of cement, 40 million tons standard coal and 30 million standard-weight cases of flat glass will be reduced. The plan has been effectively implemented in all respects, and some industries have surpassed their targets.

1.1 General information about overcapacity reduction in industries in Hebei

Hebei is a big industrial province. The outputs of its industrial products take up very high proportions of the total outputs of the whole country. The proportions of major industrial products in 2012 are shown in Table 11. Except for cement and glass, which take up relatively small proportions, the outputs of iron and steel products account for more than 20% of the total outputs of the whole country.

Table 11　Outputs of relevant industrial products in Hebei and their proportions in 2012

	Crude Steel (10,000 tons)	Steel (10,000 tons)	Pig Iron (10,000 tons)	Cement (10,000 tons)	Glass (10,000 weight cases)
National total	72,388.2	95,577.8	66,354.4	220,984.1	75,050.5
Hebei	18,048.4	20,995.2	16,350.2	12,809.8	11,382.7
Proportion	24.93%	21.97%	24.64%	5.80%	15.17%

Source: National Bureau of Statistics, Hebei Provincial Bureau of Statistics.

1.2　Current developments in overcapacity reduction in Hebei

Such large production proportions have translated into increased pressure in the area of overcapacity reduction in Hebei. For example, in 2014 overcapacity reduction in the steel, iron, cement and flat glass industries accounted for 55.56%, 55.56%, 93.29% and 72.37% of the planned tasks for the whole country, respectively①. The Hebei provincial government has attached great importance to overcapacity reduction and has attained satisfactory results.

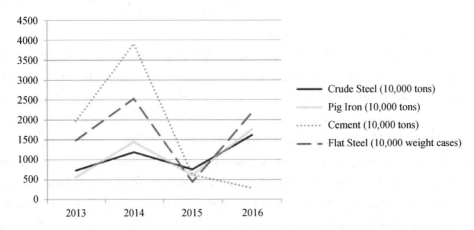

Figure 10　Completion of overcapacity reduction tasks in Hebei 2013 – 2016

As shown in Figure 10, despite some fluctuations, the average number of

---

①　Source: Website of State Council Information Office.

overcapacity reductions between 2013 and 2016 in major products remained above 1,000 units annually. The work done in 2014 was especially remarkable. In that year, the production capacities of cement and flat glass were reduced by 39.18 million tons and 25.33 million weight cases respectively, which reduced the pressures associated with the remaining overcapacity reduction tasks and significantly accelerated the adjustment in capacity layout.

During the 12th FYP period, Hebei reduced 33.91 million tons of iron production capacity, 41.06 million tons of steel production capacity, 138.34 million tons of cement production capacity, and 72.025 million weight cases of flat glass[①] respectively, achieving 68.43%, 56.51%, 230.57% and 240.08% of the original targets set out in the "6643 Program". Overcapacity reduction in cement and flat glass has already exceeded its stated goals. The reductions in steel and iron production capacities accounted for 37.26% and 43.31% of the national reductions over the same period, and were higher than the percentages of Hebei in the total outputs. Although economic growth in 2014 dropped by 1.7% year over year, the economic growth rate stayed around 6.8% over the past three years, similar to that of the remainder of the country. This means that the economic development in Hebei has been relatively stable, and without slumps. The registered unemployment in urban areas in 2016 was 3.68%, 0.34% lower than the overall level of the rest of the country. The overcapacity reduction tasks have been largely fulfilled in a satisfactory manner without large-scale unemployment or regional systematic risks.

1.3 Future plan for overcapacity reduction in Hebei

During the 13th FYP period, Hebei Province will continue to undertake 1/3 of the national reduction of iron and steel production capacity targets. According to the plan, from 2016 to 2017, 37.15 million tons of iron production capacity and 31.77 million tons of steel production capacity will be reduced. During the 13th FYP period, 49.89 million tons of iron production capacity and 49.13 million tons of steel production capacity will be reduced[②]. By the end of the 13th FYP period, the production capacities of iron, steel, cement and flat glass will be around 200 million tons, 200 million tons and 200 million weight cases, respectively.

---

① http://www.hbdrc.gov.cn/web/web/xwbd/4028818b555274660155d76169f978f6.htm.
② http://www.hbdrc.gov.cn/web/web/xwbd/4028818b55144edf015542058f042a8c.htm.

2. Analysis of cost of overcapacity reduction

Overcapacity reduction, or resolving overcapacity, refers to methods applied to transform and upgrade production facilities and products so as to reverse situations in which an industry or business cannot sell as much as it produces. Overcapacity is a example of mismatched resources, where a large amount of valuable social and natural resources are consumed in an industry with overcapacity, but fail to be fully utilized. However, overcapacity reduction is not just eliminating backward production capacity. It also concerns efforts to improve and optimize the industrial structure. Through changes in industrial production modes, overcapacity reduction is a response to the shift from an extensive economy to a circular economy, the need to reduce industrial waste pollution and damage to the environment, and the necessity of lowering negative human impacts on the climate. For Hebei, overcapacity reduction is an important part of the construction of an ecological civilization and is important for improving the living environment for the residents of the province. To achieve the goals of overcapacity reduction, governments at all levels, enterprises and other relevant stakeholders, should bear responsibility, and thus optimize and orient their functions relative to the reduction goals and invest heavily in the process of achieving such goals the whde society has paid a great price.

2.1 Theoretical framework for cost analysis of overcapacity reduction

To comprehensively estimate and calculate the cost of overcapacity reduction, efforts should be made to have a clear understanding of the connotation, external features, forms of expression and cost range of overcapacity reduction; it is also necessary to conduct cost statistic analysis and calculation in a comprehensive way, based on the actions and measures taken by the government and relevant enterprises during the process of overcapacity reduction.

2.1.1 Understanding the multiple dimensions of the cost of overcapacity reduction

In accounting, cost has different definitions relative to its application. The definition of "cost" given by American Accounting Association (AAA)'s committee on cost and standard is the value sacrificed measured by the currency that has taken place or has not taken place for a specific purpose. The definition of "cost" in Article

Report II: Cost-Benefit Analysis of Climate Public Expenditure ——A Case Study on Overcapacity Reduction in Hebei Province

2.1.2 of the CCA2101: 2005 Cost Management System – Terms, published by the China Cost Association (CCA) is: the resource[①] price paid or to be paid for the purpose of process incrementation or result effectiveness.

The problem of cost is, in a sense, a correlative and systematic problem. Certain expenditures, as a part of the total costs of overcapacity reduction, have the features of lagging and latency (For example, overcapacity reduction may result in rise in social and financial risks; to maintain a smooth operation of the economy, the fiscal authorities should account for these risks). From the accounting point of view alone, long-term invisible costs will be easily ignored, resulting in incomplete statistical results.

This report holds that the research and calculation of the cost of overcapacity reduction must be based on a theoretical basis involving inputs from economics, management and sociology. The cost of overcapacity reduction includes not only visible and measurable funding costs, but also institutional transition costs (Liu Shangxi, 2016). This report may only obtain a scientific and comprehensive view of the costs of overcapacity reduction by examining it from a variety of perspectives and dimensions, such as changes in production mode, economic and social transformation and a global perspective. Hence, the cost of overcapacity reduction defined in this report may be divided into visible cost, invisible cost and opportunity cost.

The visible cost of overcapacity reduction refers to its direct and present cost during the process, which is characterized by one-off and static natures. A "one-off" nature means that the form of funding is characterized by obvious stages. The specific scope of a one-off cost is relative clear and easy to calculate, such as the inputs in infrastructure and staff resettlement; A "static" nature means that the cost at a certain price is calculated by a certain time point.

Invisible costs refer to expenditures that are neither direct nor present during the process of overcapacity reduction, and which are characterized by frequent and dynamic natures. A "frequent" nature is the operational and maintenance cost associated with infrastructure, and also the costs linked with social management. It has a distinctive feature, namely, a rigid expenditure, and the calculation of its specific scope and amount is dynamic. Due to this "dynamic" nature, this report should, while calculating the cost, take into account any price changes within the period of time in question, including changes in overall overcapacity reduction

---

① It means all substances that may be utilized by human. The resources in an organization generally includes: human resource, material resource, financial resource and information resource, etc.

processes such as labor costs, basic construction material prices and any changes in public service costs. Practically speaking, the "dynamic" cost of overcapacity reduction in China shows a strong rising trend.

Opportunity cost refers to the value of an alternative option during the course of overcapacity reduction. It may also be understood as the value of all abandoned choices when this report chooses from any one of several programs.

Overcapacity reduction is a systematic process. It requires both cooperation from market players and appropriate interventions from the government.

2.1.2 The government's cost of overcapacity reduction

From a governmental perspective, the cost of overcapacity reduction includes both visible and measurable funding costs and institutional transition costs. For the government, overcapacity reduction itself contains comprehensive requirements for development, reform and transformation. It covers all changes in production modes, and its core purpose is to optimize the allocation of resources and facilitate smoothly all such production mode changes. Therefore, the process of overcapacity reduction cannot succeed without corresponding changes in development concepts and a deepening of the various institutional reforms which include changes in economic development modes and strategies for the protection of the environment.

The visible cost of overcapacity reduction to the government includes, but is not limited to, government expenditures on establishing financial incentives to encourage the elimination of background production capacities; expenditures to cover additional staff resettlements to prevent the social risks incurred by overcapacity reduction in enterprises; expenditures on establishing project information databases and disclosure systems so as to better grasp the progress of overcapacity reduction and ensure the process is open and transparent; social management costs incurred during the process of overcapacity reduction – the aim of such expenses is to reduce damage to the environment by lowering waste emissions from enterprises, effectively respond to climate change and monitor in real-time enterprise waste emissions (Hebei has established the only environmental protection law enforcement team in the whole country); and industrial transition costs associated with urban planning and enterprise resettlement.

The invisible costs of overcapacity reduction to the government include preferential taxes provided to enterprises reducing overcapacity. During the process of overcapacity reduction, any decision by the financial system to tighten management over lending operations in specific industries may cause economic risks, social risks, financial risks and other invisible risks such as capital chain ruptures and bankruptcies. Social risks and financial risks will also rise due to large amounts of unemployment and a rise in non-performing social asset ratios. The consequences cannot be discerned within short periods of time. When such risks take place, fiscal authorities will have to cover these costs, which are invisible expenditures.

The opportunity cost for the government in the process of overcapacity reduction includes the fact that, during such a process, the elimination of backward industries and the optimization and upgrading of growing industries will end up resulting in more inputs and fewer profits for enterprises, which will in turn lead to lower tax revenues of the government.

The main costs incurred by governments at various levels during the process of overcapacity reduction are shown in Figure 11.

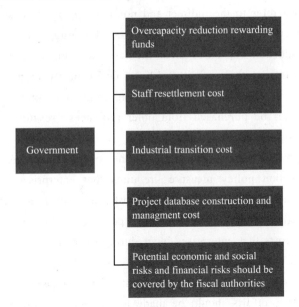

Figure 11　The government's cost of overcapacity reduction

2.1.3　Enterprises' cost of overcapacity reduction

During the process of overcapacity reduction, enterprises acting as micro players in the market will incur a range of outcomes. Some enterprises will need to be completely shut down, while others will only need to be partially shut down. For enterprises, visible costs refer to the direct and present expenditures incurred during overcapacity reduction. Some of these enterprises cannot meet the production, technology, environmental protection and quality requirements, and are thus shut down by the government due to their incapacity to limit the amount they pollute. In larger enterprises, although closed-loop production has been developed, some processes are still in need of upgrading. These enterprises continue with their operations, but will have their inefficient production capacities eventually compressed and eliminated. During this process, problems concerning staff resettlement, debt disposal and the losses from asset devaluation will occur. Through mergers, reorganizations and relocations, some of the enterprises that are shut down give play to the agglomeration effect, and thus accelerate the transformation and upgrading of their production operations. During this process, enterprises achieve industrial upgrading by investing in research and technology; in order to reduce the negative impacts of industrial waste, such enterprises invest in cleaning equipment to effectively treat waste before discharge.

Invisible costs refer to the indirect and non-current expenditures incurred during overcapacity reduction. For example, enterprises who struggle with overcapacity due to inefficient production capacities incur increased lending costs due to their being downgraded in terms of their credit standing. Overcapacity reduction results in the breaking of chains in the circular economy during production, thus requiring that some of the raw materials be purchased from other provinces, resulting in a rise in raw material, transportation and manpower costs.

The opportunity cost for an enterprise, in the context of a response to the national overcapacity reduction policy measure, requires that enterprises cannot default to whatever plan of action will result in the highest profit; they must take into account the value of rejecting or aligning with national policy.

The various costs that enterprises incur during overcapacity reduction are shown in Figure 12.

In order to estimate and calculate the cost of overcapacity reduction in a more comprehensive way, efforts should be made to classify the cost by considering the actions and measures taken by the government and enterprises during overcapacity reduction. Based on stakeholder interests and the characteristics of their activities, this report adopted the concepts of visible cost, invisible cost and opportunity cost. Thus present the cost of overcapacity reduction in the form of a matrix. In this matrix, each stakeholder involved in overcapacity reduction constitutes the vertical axis, while the categories of cost constitute the horizontal axis (see Table 12).

Report II: Cost-Benefit Analysis of Climate Public Expenditure ——A Case Study on Overcapacity Reduction in Hebei Province

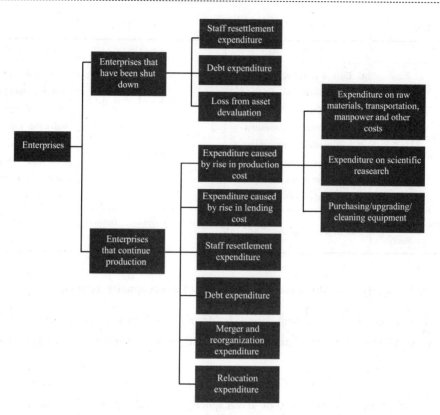

Figure 12　Enterprises' cost of overcapacity reduction

Table 12　　　　　　　Matrix of the cost of overcapacity reduction

	Visible Cost	Invisible Cost	Opportunity Cost
Government	1. Public expenditure in the form of financial incentives (cascade financial incentives) and special funds (cost of staff resettlement); 2. Additional management expenditure to establish project information databases and the disclosure system; 3. Industrial transition cost undertaken by the government.	1. Preferential taxes 2. Social security expenditure on follow-up fiscal coverage 3. Local debt risks	Reduction in fiscal income within a period of time due to the elimination of backward production capacity

Con.

	Visible Cost	Invisible Cost	Opportunity Cost
Enterprises	1. Cost of staff resettlement 2. Expenditure on merger and reorganization 3. Expenditure on relocation 4. Expenditure on scientific research and development 5. Transportation cost 6. Raw material cost	1. Increased lending cost due to credit changes 2. Increased cost caused by changes in industrial layout chains 3. Entry cost	When the market recovers, enterprises with compressed production capacities will lose market and profit opportunities, which will result in less revenues.
Financial System (e.g. banks)	Cost of due diligence	Potential systematic financial risks	

2.2 Analysis of the government's cost of overcapacity reduction

Based on the theoretical analysis framework above, combined with the overcapacity reduction data from Hebei, an analysis of the main costs to government has been undertaken.

2.2.1 Public expenditure in the form of financial incentives and special funds

To promote the reduction of overcapacity, central, provincial and municipal governments have increased investment in various ways.

The central government supports the overcapacity reduction objective of local governments through the establishment of financial incentives. On April 20, 2011, the Ministry of Finance, Ministry of Industry and Information Technology and National Energy Administration jointly issued *Administrative Measures for Central Financial Incentive Funds to Shut Down Outdated Production Facilities*, which provided incentives for the elimination of outdated production facilities during the 12th FYP period (2011 – 2015); it covered more than ten industries, including electricity, steel, coke, alcohol, and printing and dyeing. On June 14, 2016, the Ministry of Finance issued its *Administrative Measures for Special Rewards and Subsidies for Structural Adjustments of Industrial Enterprises*. In order to encourage local and central enterprises to reduce overcapacity in coal and steel as soon as possible, special rewards and subsidies, amounting to RMB100 billion (US$ 15.6 billion) were set up and provided incrementally. The amounts of the special rewards and subsidies were determined by the total size of the budget and the targets of overcapacity reduction; the special rewards and subsidies for the steel industry and the coal industry were

Report II: Cost-Benefit Analysis of Climate Public Expenditure ——A Case Study on Overcapacity Reduction in Hebei Province

determined by factors including the volume of overcapacity to be reduced, the number of workers to be resettled and the difficulty level in doing so. Proportionately more special rewards and subsidies will be provided to the provincial and central enterprises whom over-fulfill on their targets; funds will primarily be used to resettle displaced workers.

To alleviate the financial pressures associated with overcapacity reduction on enterprises, Hebei allocated rewards and subsidies to cities with districts and provincial counties (cities) in a timely manner. Also, 2014 witnessed the introduction of the *Measures for Rewards and Subsidies to Eliminate the Overcapacity of Iron and Steel Industry in Hebei Province*. RMB800 million (US$125 million) (RMB600 million for the first batch of funds) of special funds for air pollution control and RMB200 million (US$31.25 million) of special funds for the elimination of overcapacity facilities in the iron and steel industries were arranged[①]. According to the regulations, those that have removed blast furnaces of 450 cubic meters and above, converters of over 40 tons, electric furnaces of over 30 tons, in addition to the rewards and subsidies from the state, will be rewarded by Hebei with RMB250,000 (US$39,000) for every 10000 tons of iron-making capacity reduced and RMB300000 (US$47,000) for every 10,000 tons of steel-making capacity reduced. As of October 2016, RMB200 million (US$31.25 million) of provincial special funds to cut iron and steel overcapacity has been allocated.

In November 2016, Hebei Province issued its *Measures for Rewards and Subsidies to Cut the Coal Overcapacity in Hebei Province*. Subsidies will be provided based on the amount of production capacity reduced in private coal mines. The standards are calculated using two factors: first, RMB300000 (US$47,000) in subsidies will be offered for every 10,000 tons of capacity reduced; second, rewards will be provided in a meritocratic way, whereby RMB2 million (US$312,000) will be rewarded to mines with annual production capacity below 60,000 tons (including 60,000 tons), RMB2.5 million (US$390,000) to mines with annual production capacity between 60,000 tons (excluding 60,000 tons) and 90,000 tons (including 90,000 tons), and RMB3.5 million (US$547,000) to mines with annual production capacity above 90,000 tons (excluding 90,000 tons).

In November 2016, Hebei issued its *Guiding Opinions on Raising Funds through Multi-channels to Cut the Overcapacity of Iron and Steel Industry*. In accordance with

---

① http://hebei.ifeng.com/news/chengshi/ts/detail_2014_11/05/3110479_0.shtml.

the principles of the government and the voluntary involvement of the market, a compensation mechanism for reducing iron and steel capacity, subsidizing reduced capacity with stock capacity, and promoting the flow of iron and steel capacity to the enterprises with strong competitiveness was developed. So far, some of the cities and counties with more challenging overcapacity reduction targets have established rewards and issued subsidies for this end. For instance, Tangshan arranged RMB100 million (US$ 15.6 million) in financial funds in 2017 and Handan arranged RMB130 million (US$ 20.3 million) at the municipal level in 2016 to support iron and steel enterprises in their efforts to cut overcapacity.

Special funds and subsidies aim to achieve the following: allow the government to arrange a fixed proportion of the special funds for overcapacity reduction; use central special funds and local funds to support overcapacity industries undergoing capacity reduction and transformation, and offer incentives and policy support to the proportion to these enterprises who are in need of re-employing employees; provide free re-employment-related vocational skills training for laid-off workers; reallocate state-owned capital from debt restructuring to social security fund accounts to stabilize and moderately increase benefits for the unemployed.

Refer to Table 13 for the annual source of funds for overcapacity reduction from the central government and Hebei's provincial government between 2013 and 2016. As shown in the table, in recent years, overcapacity reduction funds from the central government have been increasing, and the investment volume from the central government in 2016 amounted to 15 times of that of the amount in 2013; the central government is thus an important source of public investment in overcapacity reduction.

Table 13　The annual sources of funds from the central government and Hebei provincial government between 2013 and 2016 for overcapacity reduction

	The central Government (10000 yuan)	Percentage in the general public expenditure at the central level (%)	Hebei Provincial Government (10000 yuan)	Percentage in the general public expenditure at the provincial level in Hebei (%)
2013	16,656	0.008%	10,000	0.12%
2014	119,622	0.053%	106,665	1.31%
2015	113,434	0.044%	29,776	0.35%
2016	246,227	0.090%	104,658	0.30%

Source: Hebei Provincial Department of Finance.

Report II: Cost-Benefit Analysis of Climate Public Expenditure ——A Case Study on Overcapacity Reduction in Hebei Province

> **Box: The government investment in overcapacity reduction in Tangshan City, Hebei**
>
> Tangshan is located in the eastern part of Hebei. Its GDP in 2016 reached 630.62 billion yuan (US$ 98.5 billion), with an average annual increase of 6.4%. The general public budget income amounts to 35.51 billion yuan (US$ 5.5 billion). The iron and steel output in Tangshan accounts for 2/3 of the total volume in Hebei, leaving a heavy task for overcapacity reduction. Government support for the implementation of overcapacity reduction not only includes fund investment from the governments above the provincial level, but also increasing efforts of the government at the municipal level to integrate financial capital.
>
> Overcapacity reduction in Tangshan aims to shut down outdated capacity facilities and at the same time promote industrial upgrading, with 1.04 billion yuan (US$ 162.5 million) of integrated fund at/above the municipal level of Tangshan, focusing on support for steel, cement, coking and other industries, emission reduction projects; mobilizes 1.1 billion yuan (US$ 171.9 million) at/above the municipal level to support Tangshan Best Steel Company and other enterprises to cut overcapacity in iron and steel; invests 150 million yuan (US$ 23.4 million) to support the demolition of coal-fired boilers in the central area and establishment of a centralized heating grid project; mobilizes 460 million yuan (US$ 71.9 million) at/above the municipal level to eliminate 62000 heavy-polluting vehicles; purchases 515 environmental-friendly clean energy buses; invests 225 million yuan (US$ 35.2 million) to support industrial development and establishes investment guidance funds and technology venture capital funds for industry, agriculture and service industry with the aim to create a resourceful integrated platform and boost industrial transformation and upgrading; and invests 750 million yuan (US$ 117.2 million) to support the transformation and upgrading of industrial enterprises and drive steel equipment manufacturing industries to innovate based on existing resources.
>
> In line with the air pollution control targets in Tangshan, the city has made comprehensive and planned use of air pollution control funds from the central, provincial and Tianjin governments. Tangshan got special funds of 1.25 billion yuan (US$ 195.3 million) for air pollution control, including 600 million yuan (US$ 93.7 million) in 2015 and 650 million yuan (US$ 101.6 million) in

2016. The funds have been mainly used for air pollution prevention and control tasks, such as to manage coal, reduce overcapacity in steel, promote new energy vehicles, and build capacity.

Increase the municipal capital investment. 1.14 billion yuan (US$ 178.1 million) was arranged in 2016 in Tangshan to prevent and control air pollution, of which 480 million yuan (US$ 75 million) was from 2015, and 660 million yuan (US$ 103.1 million) from 2016 as subsidies for enterprises reducing overcapacity and for the closed enterprises due to overcapacity reduction, elimination of outdated capacity and air pollution. To spur overcapacity reduction, Tangshan issued measures to provide rewards and subsidies for the capacity reduction of iron and steel, stipulating a compensation of 50,000 yuan per 10,000 tons.

Source: Tangshan Municipal Bureau of Finance.

2.2.2 Expenditure of employment stabilization subsidies

In line with the *Opinions on the Resettlement of Workers While Reducing Overcapacity in the Steel Industry to Achieve Development by Solving the Difficulties* (Ministry of Human Resources and Social Security [2016] No. 32) jointly issued by seven departments including the Ministry of Human Resources and Social Security and the National Development and Reform Commission, the General Office of Hebei Provincial Government issued its *Implementing Opinions on the Resettlement of Workers from the Iron and Steel Industry Undergoing Overcapacity Reduction*. It gives priority to the resettlement of workers and requests enterprises to take responsibility, organize work locally, and abide by relevant laws and regulations. It requires enterprises to take into consideration local conditions and make full use of market mechanisms and support measures for better resettlement work. They are also required to facilitate the resettlement of workers by expanding existing policies to stabilize employment; for example, they can include internal retirement and pending retirement, promote job transfers and entrepreneurship, encourage the merger and reorganization of preponderant enterprises, support people in dire need through services such as public welfare jobs, properly handle labor relations, carry out vocational training, and provide specific policy subsidies.

In order to better facilitate resettlement work, Hebei's provincial government intends to provide employment stabilization subsidies from its unemployment insurance fund to enterprises that pay the full amount of unemployment insurance and actively avoid layoffs. Local governments can provide one employment stabilization subsidy each year to enterprises that meet the aforementioned scope and intend to implement industrial structure adjustments and air pollution controls. The subsidy comes from the unemployment insurance fund, and will mainly cover living allowances, social insurance, job transfer training and skills training for workers. To qualify for a subsidy, an enterprise must fulfill the following: taken effective measures to stabilize jobs over the previous year without any layoffs – a subsidy can be offered equal to 50% of the total amount of unemployment insurance premiums paid by the enterprise and its employees over the same time period; taken effective measures to stabilize jobs over the previous year with its layoff rate at 1% or lower than the registered unemployment rate in the area – a subsidy can be offered equal to 40% of the total amount of unemployment insurance premiums paid by the enterprise and its employees over the same time period; taken effective measures to stabilize jobs in the previous year with its layoff rate at 1% or less than the lower the registered unemployment rate in the area – a subsidy can be offered equal to 30% of the total amount of unemployment insurance premiums paid by the enterprise and its employees over the same time period.

In 2016, Tangshan allocated RMB1.009 billion (US$157.6 million) worth of employment stabilization subsidies, of which RMB771 million (US$120.5 million) was allocated to enterprises undertaking overcapacity reduction and implementing pollution controls, accounting for 76.4% of the total. Since 2016, there has not been any large-scale unemployment caused by overcapacity reduction.

2.2.3 Industrial transfer costs

During overcapacity reduction, the adjustment and transfer of existing industrial structure and industrial docking, as well as the cultivation of new industries all face huge industrial transfer costs. In the process of industrial transfer, the costs will increase with the compensation for relevant stakeholders, resettlement of industrial transfers, and renovations of the worksites. In addition, after the transfer, there may be unsustained industrial development. Due to potential differences in the industrial structure of the new region, the interdependence of industries and lack of linkage between the upstream and downstream, it may be difficult to cultivate industry interaction and achieve the agglomeration and scale effects, which would lead to rising costs. Furthermore, there will also be costs to reconstruct public services as the new industrial area needs to re-plan traffic routes and re-construct water, electricity and other infrastructure utilities based on the needs of the workers.

2.2.4 Reduction of fiscal revenue due to overcapacity reduction

In the short term, overcapacity reduction will affect the output and profit of an enterprise. In particular, the closure of enterprises has serious impact on local economic development, resulting in reduced industrial output and tax revenue. In 2016, the gross regional domestic product of the Fengnan District of Tangshan City in Hebei reached RMB61.76 billion (US$9.65 billion), accounting for 9.8% of the GDP of Tangshan. The general public budget revenue of Fengnan District in 2016 was RMB3.15 billion (US$492.2 million), accounting for 8.9%[①] of the general public budget revenue of Tangshan. In terms of overcapacity reduction, Bainitic Steel Company from the Fengnan Town of Fengnan District got rid of 1.05 million tons of iron production capacity and 4.91 million tons of steel production capacity; its industrial output decreased by RMB2.72 billion (US$425 million) and its tax revenue dropped by RMB100 million (US$15.6 million)[②].

2.2.5 Increased financial expenditure due to higher financial and social risks

Under the influence of factors including economic downturns, market oversupply and national policies on overcapacity reduction, the industries with overcapacity are having a difficult time staying open. Some companies have suffered serious losses, and bad loans have started to increase in the banking industry. Banks are the primary driver to solve the debt crisis. If improperly handled, the results will lead to systemic financial risks.

The majority of the industries that have overcapacity are labor-intensive industries. If outdated production capacity is cut and zombie enterprises are closed, it will lead to laid-offs of a great number of people; society would thus face a need for large scale job transfers and the widespread resettlement of workers, which, if not properly handled, would lead to economic instability. For example, Tangshan Guofeng Iron and Steel Company actively responded to national policies and shut down its northern production line. 4063 workers were laid off. To ensure the timely allocation of economic compensation to their workers, the company raised RMB300 million (US$46.9 million) from various channels to resettle the laid-off workers. If the utilization of these funds are inefficient, the laid off workers will suffer, the local employment market will become unstable, and costs to the government will increase.

2.3 Analysis of enterprises' cost of overcapacity reduction

2.3.1 Increase in resettlement costs for workers

As progress is made in overcapacity reduction, the question of worker

---

① Source: Tangshan Statistic Brief 2016.
② Provided by the Finance Bureau of Fengnan District during the survey.

resettlement is becoming more complex and challenging. Based on pressures from rising labor costs and a significant gap in the resettlement fund, there is risk of recessive unemployment, examples being internal retirement and rotating holidays. When in the process of resettling workers, enterprises should follow the Labor Law and Labor Contract Law strictly, to improve the labor relationship and pay economic compensation. Surveys show that in 2016, Hebei resettled a total number of 57,785 workers who were displaced due to overcapacity reduction; 32,450 of these people transferred jobs, 4,755 were internally retired, 18,295 people terminated their labor contracts, and 2,285 people received natural attrition. It is estimated that it takes an enterprise RMB90,000 – 200,000 (US$ 14,000 – 31,000) to resettle one employee, thereby confronting enterprises needing to engage in employee resettlement with large financial obligations.

2.3.2 *Increase of production and operation costs*

Overcapacity reduction increases the operational costs of downstream enterprises. After 2016, most of the iron and steel enterprises in Hebei were joint production enterprises. When production at the front-end was reduced, enterprises at the back-end needed to outsource, which directly increased operating costs. This is mainly reflected in the fact that with the expansion of overcapacity reduction, the cement and coke industries in Hebei could not meet the needs of the province, and thereby decided to purchase from other provinces, a decision which increased transport costs significantly.

Overcapacity reductions also increase an enterprise's costs in terms of technological transformation, and research and development. Overcapacity reduction is not simply the act of shutting down outdated capacity, but includes the need to optimize existing capacity structures and protect the environment. In turn, increased investment by an enterprise in technological research and development, safe production, environmental protection, energy saving and emission reduction will result in increased production costs. For instance, iron and steel companies in the Fengnan District of Tangshan City in Hebei invested a total of more than RMB2 billion (US$ 312.5 million), and completed 152 atmospheric upgrading projects, including desulfurization and dust removal; they invested in 2016 RMB200 million (US$ 31.25 million) and completed a total of 19 treatment projects on sintering machines to ensure that emission concentrations from sintering machines, shaft furnaces for desulfurization and particulate matter from flue gas meet the national special emissions standards.

2.3.3 Increased cost for corporate financing

Overcapacity reduction has increased the financing costs for enterprises. With the nationwide implementation of overcapacity reduction policies, banks have started to exclude overcapacity enterprises from their support, thereby tightening the credit scale for these enterprises, and increasing the cost and difficulty for enterprises to raise funds. Even leading enterprises may be affected and unable to obtain funding. When banks stop releasing loans to enterprises or call in loans ahead of schedule, it creates serious problems. A survey in Hebei showed that in Handan alone, RMB7.5 billion (US$1.17 billion) of loans were withdrawn by the banks. Without loans from banks, sourcing financing becomes more complex and expensive.

2.3.4 Increasing difficulties in debt management

By cutting excess capacity, enterprises face greater operating pressures, substantial declines in benefits, higher operational risks, and even, in the worst cases, the suspension of production altogether. It is easy for these enterprises to become tangled in a debt. Enterprises will be greatly affected by such dire circumstances, resulting in adverse outcomes such as unpaid wages and delays in the payout of social security insurance. If this isn't bad enough, it becomes even more difficult to manage debt after overcapacity reduction. Take Tangshan City as an example. In 2017, four steel enterprises will be closed, with a total liability of RMB5.4 billion (US$0.84 billion). Among the liability, RMB980 million (US$153.1 million) are bank loans, with the rest being a mix of corporate loans, social funds, arrears, arrears of wages and social insurance.

2.3.5 Relocation cost

Hebei Province sped up its major relocation and renovation projects by relocating its key polluting industries – iron and steel coking – from urban areas to industrial parks, in accordance with the 1:1.25 ratio of capacity reduction replacement. For the six major projects – Construction of Shougang Jingtang II, Relocation of Tangshan Bohai Iron and Steel Company, Relocation of Shijiazhuang Iron and Steel Company, as well as the Relocation of Yongyang Special-steel Company, Taihang Iron and Steel Company and Ji'nan Iron and Steel Company from Urban Areas to Industrial Parks – that are listed in the Restructure Plan for Iron and Steel Industry in Hebei, a special working group was set up to provide solutions and enable the projects to be completed as early as possible. A comprehensive exit plan has been introduced to Xuanhua Iron and Steel Company, which merges it with parts of Tangshan Iron and Steel Company and Chengde Iron and Steel Company, thereby reducing its capacity and relocating it to a new area. In this process, there will be relocation costs on the enterprises.

2.3.6　*Merger and reorganization cost*

After overcapacity reduction, some enterprises cannot continue normal operations. After an initial survey, there were 11 "zombie enterprises" in the iron and steel industry in Hebei. Taking into account the interests of different stakeholders, including the workers, such enterprises engaged in mergers, restructuring, bankruptcy, and other approaches. In the area of mergers and reorganization, enterprises needed to pay capital investment costs to engage such approaches.

2.3.7　*Opportunity cost*

The opportunity cost of an enterprise includes part of the measures to reduce excess capacity, including production suspension and limitation. But periodic production suspension will not only affect the income of the enterprise, but also increase labor costs and corporate maintenance costs. Hebei sets strict controls over the introduction of new capacity. In the case of market recoveries, the loss of profit opportunities after capacity reduction has led to declines in business income.

2.4　Analysis of cost of overcapacity reduction in the financial system (with banks as an example)

2.4.1　*The cost of the rising bad debt rate*

Statistics shows that the average debt ratio for iron and steel enterprises in Hebei in 2016 was 65%; this figure will likely rise in 2017. Among them, loans from banks and financial institutions account for a considerable proportion. In 2017, four iron and steel enterprises in Tangshan City will go bankrupt with a total liability of RMB5.4 billion (US$ 0.84 billion), of which RMB980 million (US$ 153.1 million) is bank loans. Production capacity will cause banks to increase their percentage of non-performing loans. Among the industries with excess capacity, inefficient enterprises with high debt take up a lot of credit resources. Once the capital chain breaks, it will result in non-performing loans that the banks will have to endure; this is an unfavorable situation, given that such a rise in non-performing loans means a reduction in the the amount of loans, thus forming a vicious cycle that could produce systemic risks.

2.4.2　*Increased cost of due diligence for certain industries and enterprises*

Fearing that overcapacity reduction will turn questionable debts into bad debts and cause an increase in non-performing assets, banks have begun producing lists of the industries targeted for capacity reduction, and either tightened their requirements for extending credit to them or stopped lending altogether. In addition, in order to prevent and manage risks, banks have enhanced due diligence for industries undergoing overcapacity reduction, which has resulted in increased costs. A survey in Hebei showed that in Handan alone, banks have reduced the amount of available credit by about RMB7.5 billion(US$1.17 billion).

3. Analysis of the benefits of overcapacity reduction

At first glance, overcapacity reduction is only about cutting excess capacity. But it also involves the optimization of production structures, reduction of industrial emissions, improvement of resource utilization efficiency, and responses to climate change. It covers a wide range of industries and sectors; its performance has complex and non-market qualities. Benefit analysis related to overcapacity reduction from only one perspective is bound to be biased; thus multiple aspects should be considered to assess the efficiency of such reduction. this report intends to evaluate the benefits of public expenditures on overcapacity reduction from three aspects: ecological sustainability, social sustainability and economic sustainability. In terms of the specific evaluation indicators, quantitative data is mainly used to analyze the effectiveness of overcapacity reduction in Hebei, which is then compared with the national data. For the fields where data is unavailable in the short-term, a qualitative analysis has been done.

Table 14　　Benefit analysis framework of overcapacity reduction

Type of Benefit	Main Reflection
Ecological sustainability	Reduce resource consumption and emissions
	Improve the quality of ecological environment
	Promote the construction of ecological civilization
Social sustainability	Promote the health of residents
	Promote the development of people's livelihood
Economic sustainability	Enhance the quality of economic development
	Promote economic transformation and upgrading
	Optimize the industrial layout

Report II: Cost-Benefit Analysis of Climate Public Expenditure ——A Case Study on Overcapacity Reduction in Hebei Province

3.1 Eco-environmental benefits

The core objective of overcapacity reduction is to reduce the current over-reliance on resources and labor inputs for economic development and to improve the way the Chinese economy approaches sustainability objectives. In terms of the resource use efficiency, overcapacity reduction involves the reduction of resource consumption and emissions and the improvement of air quality. Regarding ecological and environmental benefits, overcapacity reduction is crucial for building an ecological civilization.

3.1.1 Reduce resource consumption and emissions

For a long time, China was an underdeveloped country with vast resources. But mass resource exploitation paired with new technologies accelerated the consumption of resources to unforeseen heights. In the case of excess capacity, resource consumption does more harm than good. On the one hand, people are consuming the resources of future generations. On the other hand, such consumption has devastated China's environment, from air to water to soil. When environmental endurance reaches a certain threshold, all further environmental damage will be for intents and purposes irrevocable. Overcapacity reduction cuts inefficient production capacity through resource utilization. When the traditional capacity is reduced, new capacity will have been created with sustainability in mind, thus ensuring the rational use of resources.

In Hebei, during June 2016, the Hebei provincial government revised six local standards dealing with environmental protection, energy consumption, water consumption, quality, technology and safety in order to push overcapacity reduction; these local standards were stricter than the national standards or industry average. Judging from the current situation, energy consumption in Hebei has decreased significantly, and the energy consumption indicators and the added value of every RMB10,000 for industrial enterprises are both faster than the national level.

Industrial energy consumption has been declining relatively fast in Hebei over recent years. Except in 2016, the decline rate is higher than the national average (see Figure 13 for details).

In terms of water consumption, for every RMB10,000 of added value, Hebei has maintained a high water consumption efficiency. Its water consumption in 2014 was only 29.69% of the national level, and its water consumption has been on rapid decline since. In 2013, the water consumption of every RMB10,000 yuan of added value already ranked fourth in the country, only after Tianjin, Shandong and Beijing.

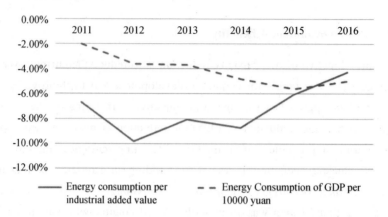

Figure 13 The comparison of energy consumption between Hebei and the country in 2011 – 2016

Source: Annual statistic reports of Hebei, Annual reports of National Bureau of Statistics.

In 2016, the Hebei Water Resources Department issued more stringent water management objectives, and planned to reduce the number to 12 tons by 2020 (see Figure 14 for details).

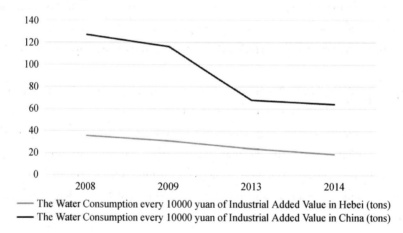

Figure 14 Change of water consumption every 10000 yuan of industrial added value in Hebei in recent years

Source: NDRC Website, National Statistic Reports, Hebei Daily, Hebei Water Conversation Plan.

3.1.2 Improving the quality of the ecological environment

A good ecological environment is the fairest public product and the most inclusive benefit for the general public. At present, however, the environment is a weak point for China, and so addressing it has become an important part of the current economic and social system reform. Reducing overcapacity is a process whereby emissions are reduced through structural reform, technology adoption, engineering innovation, and the development of more sustainability – focused management strategies. It is a holistic process, involving not only the enterprise itself, but the government and the local community.

Hebei has also made improvement in the area of air quality. The annual average concentration of PM2.5 has fallen by nearly $40\mu g/m^3$ during the period of 2013 – 2016, down more than 30%, with the average annual reduction exceeding 10%. Other major pollutants, such as PM10 and ozone, have also seen significant reductions in their concentrations. In 2016, excellent or good air quality was reported for 75 days more than in 2013; heavy pollution was reported for 50 days less. Generally speaking, the quality of the air has improved significantly during the period of 2013 – 2016 (see Figure 15 for details).

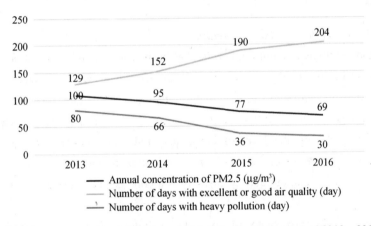

Figure 15 Air quality indicators of Hebei during the period of 2013 – 2016

Source: provincial environmental status bulletins over the years.

3.1.3 Promoting the development of an ecological civilization

Since the 18th National Congress of the CPC in 2012, President Xi Jinping has integrated the concept of an ecological civilization into China's economic, political, cultural and social development. The People's Government of Hebei has accelerated the building of an ecological civilization since 2015. Reducing overcapacity is an important aspect of this effort. Reducing overcapacity means to change the existing development approach, which wastes resources indiscriminately, to reduce dependence on resources and labor, to accelerate away from traditional industry made characterized by high pollution, high consumption, high risk, low benefits and low output, and to regard green, eco-friendly processes as the only way to achieve green development. Structural optimization is also employed to promote the transformation and upgrading of the entire economy, and to achieve the synchronization of industrialization, informatization, urbanization, agricultural modernization and green development. In the long run, the efforts to reduce overcapacity are conducive to promoting the development of an ecological civilization.

3.2 Social benefits analysis

In the long term, reducing overcapacity is conducive to sustainable social development. It can promote the health of residents by squeezing out inefficient, surplus production capacity, promoting the upgrading of production technologies and improving environmental quality. Through the responsible use of resources, it also contributes to the long-term stability of employment, incomes and people's livelihoods.

3.2.1 Promoting residents' health

A good ecological environment is the basis for human survival and health. Without good health, there will never be prosperity for the whole society.

The extensive economic growth pattern that has prevailing for some time has negatively impacted the health of Chinese society. By raising energy consumption standards and reducing the level of emissions, reducing overcapacity can curb the exacerbation of environmental decay. The elimination of inefficient capacity and the development of efficient and quality capacity is a core aspect of the green development concept.

A look at the provincial picture of respiratory diseases (tuberculosis and influenza) over the past two years reveals that both the respective number of cases and the share of their respective cases to the total cases of all diseases have shown relatively strong seasonality, small variation from extreme values and minimal year-on-year differences in the years concerned. There might be a relationship between overcapacity reduction and the health status of residents. However, this will be a long-term effect which cannot be observed yet.

3.2.2 Improving people's livelihood

In the short term, Hebei Province has given top priority to protecting worker rights and ensuring that displaced workers find adequate resettlement. It has widened the options for relocating laid-off workers, including internal diversion/transfer, entrepreneurship, early retirement, and transfers to other public offices. In total,

Report II: Cost-Benefit Analysis of Climate Public Expenditure ——A Case Study on Overcapacity Reduction in Hebei Province

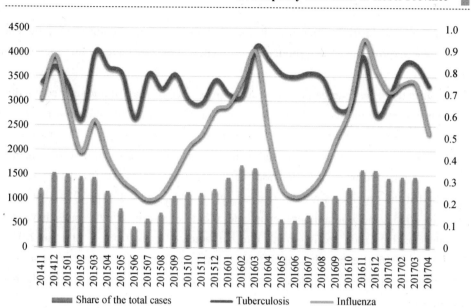

Figure 16 Number of cases and share of the total cases of the two respiratory diseases in Hebei in recent years

Source: website of the Provincial Health and Family Planning Commission of Hebei.

Hebei accommodated 57,785 workers in 2016. Government expenditures in this area have effectively prevented adverse social impact through unemployment. In the long term, the campaign to eliminate outmoded overcapacity can lead to the responsible use of resources, promotion of a clean environment and sustainable social development.

3.3 Economic benefits analysis

The efforts to reduce overcapacity are an important part of supply-side structural reform in China. Under the new normal, the original growth pattern has been shown to be entirely deficient, hence the necessity to adjust production structures through supply-side structure reform. The campaign to reduce overcapacity has improved the quality of economic growth, promoted a new approach to economic development and facilitated the optimization of industry distribution.

3.3.1 Improving the quality of economic growth

The campaign to reduce overcapacity can improve short-term control policies and ensure sustainable economic growth. By directly adjusting the capacity structures, it can restrict the huge input of production materials and prevent price competition due to excess supplies of the industrial products. Moreover, by directly guiding the adjustment of industrial structures, it can contribute to the reduction of production costs, the improvement of production efficiency and the increase of added value inputs.

Currently, the campaign to reduce overcapacity has improved competition between industries in Hebei. In recent years, provincial data on industrial enterprises above the designated size have outperformed the national averages. This means that the province has preliminarily adapted to the new normal of economic development in China. A comparison of the industrial data of enterprises above the designated size reveals that the province's growth rate has rebounded from 4.4% in 2016, the lowest level in 2015; the operating revenue and the total profit growth rates have both become positive; total profit growth rate has bounced back from -11% in 2015 to 18.9% in 2016, seeing the highest level over the past five years. Overall, the industrial sector above the designated size in Hebei has seen its profits increase dramatically.

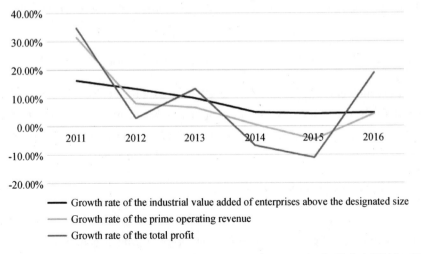

Figure 17 Indicators of industrial enterprises above designated size in Hebei (2011 – 2016)

Source: website of the Provincial Bureau of Statistics of Hebei.

3.3.2 Promoting economic transformation and upgrading

Economic transformation and upgrading are important objectives in the campaign to reduce overcapacity. Hebei has improved in this area through its capacity reduction efforts.

On the one hand, Hebei has improved the technology associated with its capacity. Based on its local energy standards, it encourages enterprises to increase

Report II: Cost-Benefit Analysis of Climate Public Expenditure ——A Case Study on Overcapacity Reduction in Hebei Province

R&D in iron and steel production, equipment adoption and technology development; it wants enterprises to conserve energy while producing high-end iron and steel products, thus giving them the opportunity to be included in the national catalogue of high-tech enterprises. It is also striving to foster steel – consuming enterprises that manufacture construction-purposed steel structures or metal products and develop non-steel enterprises that process and distribute steels. By the end of the 13th FYP period, the ultimate objective is to make the equipment of the province's iron and steel industry internationally advanced, raise the share and variety of special steels, and exceed 30% in operating revenue from non-steel businesses.

On the other hand, Hebei is trying to improve the quality of its incremental capacity. It is guiding the iron and steel manufacturing industry in integrating internet technologies to increase efficiency; a core focus concerns developing the transport equipment, energy equipment, engineering and special equipment, and basic parts industries and ensuring that the value add from the equipment manufacturing industry exceeds that of the iron and steel industry over the next year. In addition, the province has exerted great efforts to develop new technologies, industries, businesses and models by executing a three-year action plan; it includes key benchmarks for developing a modern service industry, a high-tech industry amplification plan and a small-and medium-sized sci-tech enterprise growth plan. Hoping to make the value add from the service industry account for 45% of the total GDP and the value add from strategic, emerging industries account for 20% by 2020.

Hebei has adjusted the structure of its primary, secondary and tertiary industries. In 2016, the share of its secondary industry dropped 16.8%, while the share of the tertiary industry increased 17.8%, from 2011.

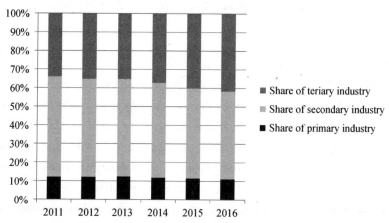

Figure 18 Share of the value added of the primary, secondary and tertiary industry in Hebei

According to the situation of existing sci-tech industries, Hebei's measures to transform the relevant industrial structures has been effective. The growth rate of the sci-tech industries was higher than that of the industrial enterprises during the period of 2011 – 2016. The gap between the two narrowed in the early stage of that period, but as the former remained above 10% and the latter began to fall, the gap between the two gradually widened. In the past three years, the growth rate of the sci-tech industries remained 8 percentage higher than that of industrial enterprises. Accordingly, the contribution of the sci-tech industries to the regional economy has been increasing consistently. By 2016, it already accounted for 18.4% of the value add from the industrial enterprises, 2.2 percentage higher than that in 2015. The effectiveness of such economic transformation is therefore clear.

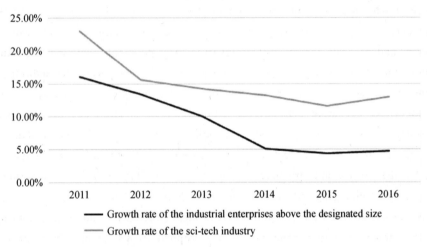

Figure 19 The sci-tech industry grew faster than the industrial enterprises above the designated size during the period of 2011 – 2016

Source: website of the Provincial Bureau of Statistics of Hebei.

3.3.3 Optimizing the industrial layout

Industrial layouts exist to optimize the regional economic structures and spatial organization structures of the industries through the responsible allocation and flow of industrial factors. The optimization of the industrial layout is conducive not only to the clustering of industries, formation of upstream and downstream industries and relocation of industrial activity from ecologically vulnerable areas to areas with relatively high carrying capacity, but also reducing pollution, conserving ecological resources and optimizing economic structures. Responsible industrial layouts can help achieve good social, economic and ecological effects, and are therefore key for harmonizing economic, social and environmental development.

Hebei's industrial layout is confronted with problems such as scattered distribution, low level integration, horizontal divisions of work and inadequate coordination between industry clusters. Irresponsible industrial layouts have restricted the further development of the province.

So Hebei plans to combine the campaign to reduce overcapacity with efforts to optimize its industrial layouts. Such a combination aims to address the low concentration and irresponsible geographic distribution of the iron and steel industry through consolidation and restructuring, layout optimization and outward transfer; such a process would raise the level of development and sharpen the competitive edge of the iron and steel industry.

Concerning consolidation and restructuring, Hebei will combine the efforts of key enterprises with the government to undertake cross-regional and cross-industrial downsizing and restructuring in order to raise the level of industrial concentration. By the end of the 13th FYP period, the number of iron and steel enterprises in the province will be reduced to 60 from 109 and led by such flagship enterprises as HBIS Group and Shougang Group; it will also consist of 3 local groups and 10 characteristic enterprises in addition to the flagships.

Concerning the optimization of industrial layouts, Hebei will cut the capacity of environmentally sensitive areas and the areas on the rim of Beijing and Tianjin to encourage and guide the iron and steel production capacity to relocate to coastal port areas or at least move away from inner city areas into the industrial parks. During the 13th FYP period, the cities of Zhangjiakou, Baoding and Langfang will have all of their iron and steel production capacity cut; Qinhuangdao and Chengde will have 50% of their capacity cut; and other cities and their surrounding areas will also gradually dismantle their steel plants or relocate them to industrial parks or coastal areas.

Concerning international cooperation in this effort, Hebei will encourage leading enterprises to set up production and manufacturing bases overseas to handle key projects. During the 12th FYP period, the province has realized rapid development in go-global efforts and international capacity cooperation. It has accumulated US$ 7.05 billion worth of direct foreign investment, which was 4.7 times that of the 11th FYP period. Particularly, a number of international capacity cooperation projects, such as the 2.2 million ton steel project of Hegang Group in Serbia and the 1.2 million ton cement project of JIDD, have been successfully implemented.

4. Key conclusions and recommendations

4.1 Key conclusions

4.1.1 Overcapacity reduction is a significant measure for dealing with climate change; more research should be done in this area.

Paris Agreement, aimed at stepping up the enforcement of the UNFCCC, presents the global blueprint and visions for tackling climate change and achieving green economy after 2020; such an achievement would be a notable landmark in the history of climate governance. China has assumed environmental responsibilities and is actively working towards its Paris Agreement commitments. China has clarified a series of targets, including peaking its CO_2 emission by 2030. These targets have been integrated into China's overall development agenda. Now and henceforth, the promotion of low-carbon models of economic development is critical necessary to close outdated production facilities and promote the transformation of economic development patterns, improve the quality of economic development and cope with climate change. It is of utmost necessity to strengthen research on institutional arrangement, policies, tools and cost-effectiveness assessment related to overcapacity reduction.

4.1.2 The exorbitant cost of the campaign to reduce overcapacity is to be shared by the government and enterprises.

To more comprehensively evaluate the cost of overcapacity reduction, there is a need to take into account the actions and measures of government, financial institutions and enterprises and to divide the cost among all stakeholders accordingly. Considering all the stakeholders and their respective characteristics, this report used visible costs, invisible costs and opportunity costs, and constructed a cost matrix. Our multi-dimensional matrix analysis shows that stakeholders like the government and banks have paid a high price to reduce overcapacity. In our analysis of the overall cost, apart from the explicit expenditures of the central government, and given that the provincial government and the county/city governments supported capacity reduction efforts with earmarked funds, this report also paid attention to the implicit expenditures of the governments, banks and enterprises, although the latter is hard to quantify.

4.1.3 The campaign to reduce overcapacity has generated preliminary effects in economic, social and ecological sustainability.

The campaign to reduce overcapacity aims to achieve a shift from an extensive economy to a circular economy by means of reforming the mode of production in the industrial sector, thereby reducing environmental pollution, damage due to industrial wastes and the negative impact of human activities on the climate. For Hebei, the effort to reduce overcapacity is an important component in the effort to build an ecological civilization. By 2016, overcapacity reduction has generated initial effects in social, economic and ecological sustainability. In terms of the social benefits, it has been associated with improvement in long-term employment stability, income increase and an overall improvement in the livelihood of local residents. In terms of economic benefits, it has contributed to the improvement of the quality of economic growth and economic transformation, and to the optimization of industrial structures and layouts.

4.1.4 The cost is higher than the benefits of the campaign to reduce overcapacity in the short term.

Based on field investigations and the cost-benefit analyses in this report, this report conclude that the cost of the campaign to reduce overcapacity is higher than the benefits in the short run. Stakeholders, including the government, enterprises and banks, are under considerable pressure from current expenditures. Analysis on the debt, cost and profit data of the province's industries indicate an increase in the current cost of production and a decrease in the profitability. However, this report argues that the benefit is ultimately higher than the cost in the long run. At the micro level, the campaign has already helped enterprises reduce cost and increase profit. By improving enterprise performance in areas such as environmental protection, energy consumption, quality and safety, it is able to effectively eliminate outmoded capacity and lower the overall production cost to society. By adjusting the structure of production and increasing the added value of production, it can drive a steady increase of profit in the end. At the macro level, as elucidated above, the campaign has a great potential in helping achieve social, economic and environmental objectives.

4.2 Recommendations

4.2.1 The campaign to reduce overcapacity needs to be put in the context of developing a circular economy and optimizing industrial layouts.

In view of our field investigation in Hebei, this report suggests that the campaign to reduce overcapacity should not only focus on emission reduction; instead, it should take into consideration the need to develop a circular economy, recycle waste, and adjust and optimize industry structures in policy design.

4.2.2 There is the need to implement policies in a differentiated manner and to optimize policies in a dynamic manner.

This report must adjust the tasks and goals of overcapacity reduction in a timely and optimized way. The policies aimed to reduce overcapacity must be implemented in a differentiated manner, with industry chains and supports, taken into consideration. For example, policies may be enforced with respect to the single-product coking industry, but may be adjusted with respect to enterprises that integrate coking and steel production and whom require the installation of desulfurization equipment. Differentiated treatment may be provided for different products. For example, many aluminum products in the electrolytic aluminum industry, and many special steels in the iron and steel industries, are all products that fall short of demand; in light of this, policies should guide and support such industries concerned to improve their technology and sharpen the competitive edge of their products.

4.2.3 The incentive and subsidy policies of the central government need to be perfected.

To support local efforts to reduce the overcapacity of the iron, steel and coal industries, the central government has set up a special incentive and subsidy fund for industrial enterprises to divert and accommodate workers in the process of reducing excess capacity. This report learnt from the field investigation that there was still room for improvement in the area of subsidies and incentives. For example, the measures for allocating the national incentive and subsidy fund need to be adjusted; the flexibility and effectiveness of the fund's uses need to be improved. The allocation of the fund should take into consideration the overall tasks of enterprises, the number of workers in need of resettlement and the difficulties associated with reducing overcapacity. So, enterprises may use the fund for a broad range of purposes such as capacity compensation, worker relocation, compensation for asset loss, debt repayment and the internal transfer of workers. Under the current provisions, there is at least the need to clarify that the fund can be used not only for worker resettlement purpose, but also to cover related costs such as worker endowment insurance, payment of the premiums in arrears for social insurance. In transferring funds, not only workers directly affected by capacity reduction activities, such as the closure of equipment, but also workers who are indirectly affected, such as those who work in industries that manufacture now obsolete machinery, need to be considered. Also, enterprises who decided to close earlier than scheduled need to be supported. Moreover, performance evaluations of the incentive and subsidy policies need to be improved, so that the policies are result-oriented and their incentive role is strengthened.

4.2.4 The analytical framework for climate public expenditure review needs to be expanded and relevant research needs to be strengthened.

The cost-benefit analysis on the climate public expenditure project is an important part of climate public expenditure evaluation, and is of great significance for improving the performance of public expenditures and responding to climate change. This report has established a multi-dimensional matrix analytical framework and systematically analyzed the costs and benefits of overcapacity reduction. With respect to the cost analysis, it distinguished between stakeholders, such as the government, banks and enterprises, and divided costs into visible cost, invisible cost and opportunity cost. With respect to the benefit analysis, it makes macro and micro analyses in terms of the ecological benefits, the social benefits and the economic benefits. The multi-dimensional matrix analytical framework this report proposes needs to be improved and expanded via more case studies.

4.2.5 The collection and statistical analysis of information and data relating to climate public expenditure needs to be enhanced.

The cost-benefit analysis for climate public expenditure is confronted with not only technical challenges, but also challenges from data availability. As evaluation of climate public expenditure is a completely new domain, the statistical scope and criteria for climate public expenditure have yet to be unified, and the evaluation method has yet to be perfected. Climate public expenditure involves many departments, so the generation of basic cost-benefit data requires cooperation on the part of a variety of government agencies. In the future, there is a need to enhance the collection and statistical analysis of basic climate public expenditure information, accelerate the development of a national climate change database and sort out the data and information resources scattered across the development and reform, fiscal, environmental protection, agricultural, health and family planning, and statistical departments, to achieve interconnection, sharing, updating, integration and ultimately a synthesis of the information on climate public expenditure.

Conclusion

As the world steps into the development era led by the SDGs, green development has been prioritized in China's most recent social-economic development planning – the 13th FYP. To achieve this, China's vision of an ecological civilization is proposed as the overall framework to consolidate policy tools that direct resources and efforts to promote a balanced and inclusive economic growth which does not compromise environmental sustainability.

Against this backdrop, CPEIR II was launched to provide institutional, policy and public expenditure analysis at the sub-national level to systematically review the current status of multi-dimensional support for climate change activities and gauge the level of complementarity among different elements. The goal is to identify key areas for further intervention and provide possible solutions to strengthen climate change governance in China.

The report focuses on Hebei – a heavily industrialized place with serious environmental problems. The province represents a pertinent case to test out means to realize the ecological civilization envisioned by China, given its challenge to manage structural economic transition and its green growth ambitions. Its experience in this regard could provide valuable lessons for other provinces, or sub-national jurisdictions in other developing countries, which are exploring their own strategies for curbing climate change and protecting environmental integrity while maintaining strong economic growth.

1. Climate public expenditure and institutional review

Hebei is making remarkable progress in addressing climate change given the strategic importance it has attached to green growth. At the institutional level, in 2008 it set up an inter – departmental coordination mechanism, namely, the Leading Task

Force on Tackling Climate Change in Hebei (LTTFTCC), to take charge of climate change activities. This has ensured that the various ministries involved are simultaneously engaged with tasks that help reduce climate change in different ways. Broadly speaking, different lines of ministries are engaged according to a range of functions they could perform, including climate mitigation and adaptation, technology development and capacity building, as well as international cooperation.

Under the overall leadership of the LTTFTCC, Hebei has issued more than 50 policy documents to address climate change, focusing predominantly on the mitigation of CO_2 emission (e.g. by improving energy efficiency, optimizing energy structures, and increasing forest carbon) and adaptation (e.g. through capacity building in water resource management, agriculture and forestry, marine resource and infrastructure, and disaster risk management). These documents are primarily issued to guide domestic efforts.

To implement the policies formulated, various sources of financial capital need to be mobilized to provide support. For instance, fiscal funds are primarily distributed in nine main fields that contribute to tackling climate change, ranging from ecology restoration, air pollution control, water management, and energy efficiency to agricultural development. It is worth mentioning that many innovative financing approaches have been adopted to address climate change issues (e.g. Public-Private-Partnerships (PPP)), implying that an expanding array of sources are being applied to complement fiscal funds and enlarge the pool of climate finance.

This report has discovered that public expenditure on climate change was on the rise during the 12th FYP period. In 2011, around 8% of the provincial budget was spent on high or moderately related climate change activities. This number kept growing year over year until 2015, when about 11% was recorded, equivalent to RMB39 billion (US$ 6.1 billion) of fiscal expenditure. Moreover, it is worth noting that this level of climate public expenditure was above spending at the central level. For example, around 10% of Hebei's budget was allocated for high or moderately relevant climate change activities compared to that of 7% at the central level.

2. Cost-benefit analysis

This report also conducted cost-benefit analysis associated with overcapacity reduction in Hebei. An overarching framework has been established to examine costs and benefits in a multitude of dimensions. Specifically, the visible, invisible and opportunity costs for each stakeholder are defined and identified in the report.

As to the state, the main costs refer to the reduction of tax revenues and provision of subsidies to the enterprises for the resettlement of the laid-off. The latter alone cost Hebei RMB105 million in 2016. As to the enterprises which are assigned the task of reducing overcapacity, the main costs relate to staff replacement and industrial upgrading, among others. It is estimated that to resettle one worker costs RMB90 – 200,000 for a company. In 2016, 57,785 laid-off workers were resettled, resulting in RMB5.2 million – 11.6 million (US$ 0.8 million – 1.8 million).

Concerning benefits, positive effects are diagnosed in all three pillars of sustainability in Hebei: energy consumption levels of industrial enterprises has kept decreasing; economic structures are transitioning and being optimized (e.g. the share of tertiary industry in gross GDP increased by 17.8% in 2016 compared to that of 2011); and better air quality is likely to result in better health for local residents (although the effect may not be manifested in the short term).

3. The next step

This report has attempted to review climate institutions, public spending and its cost-benefits at the sub-national level in China. There are two important methodological contributions of the report apart from empirical findings in Hebei. First, when categorizing the climate-relevance of budget items, the it considers the nature of the activities; i.e. whether it mitigates or adapts to climate change. This break-down could help better understand the potential effects of climate spending. Further, it could help broadly track the portfolio of activities climate spending is targeted at. This could provide useful insights for decision makers to introduce, adjust or balance budget items in order to maximize their intended outcomes. For instance, more of the budget may be reserved for highly relevant activities which could simultaneously contribute to mitigating and adapting to climate change.

Second, the report has tentatively established a holistic theoretical framework to guide the cost – benefit analysis of climate spending. This is one of the first attempts to set up a multi-criteria framework to assess costs and benefits across all pillars of sustainability, thereby raising the challenge of balancing different development priorities (e.g. economic growth, social protection and environmental protection) with the interests of different stakeholders. Moreover, the framework has directed attention to the importance of a systematic approach, which places great emphasis on the inter-connections between industries and the vertical development within specific industries. For instance, when costs are calculated, it is indispensable to consider the

Conclusion

supply and demand of goods and services along the entire industrial chain that climate activity could have potential bearings on.

This report has laid the foundation for further research on a few topics. First, the cost – effectiveness of climate public spending; i. e. the 'value for money' tradeoff can be investigated going forward once more data is gathered. Second, more domestic financial flows can be analyzed and consolidated in preparation for building up an integrated national development financing framework. In China's context, it is of particular importance to broaden the analysis in connection with green finance, within which a wider range of financial flows (e. g. green bonds, insurance, emissions trading schemes) can be leveraged to promote green development, including addressing climate change. During this process, the private sector must play a crucial role. However, their financial contributions need more stringent monitoring and evaluation to ensure development effectiveness.

Annex I. Policies on climate public expenditure

This part analyzes policies on climate public expenditure and innovative expenditure patterns of Hebei. Fiscal expenditure information is divided into two categories: financial investment policies and innovative expenditure patterns. Financial investment policies can be divided into 9 categories: direct response to climate change, ecological restoration, groundwater over-exploitation management, energy saving and emission reduction, and energy structure optimization, energy efficiency improvement, comprehensive agricultural development, rural transformation, flood prevention and disaster relief. Innovative expenditure patterns can be divided into 4 categories: public-private partnership, international loans, clean development funds and emission trading.

1. Financial investment policies

(1) Direct response to climate change

Between 2013 and 2015, Hebei Province invested a total amount of RMB 45.73 billion (US$ 7.15 billion) on air pollution control (RMB 12 billion (US$ 1.9 billion), 16.03 billion (US$ 2.5 billion) and 17.7 billion (US$ 2.8 billion) respectively), and continued to innovate in fund allocation, supervision and evaluation mechanism, which ensured the smooth implementation of pollution control projects. Through budget integration and adjustment, coordinate the operating budget of special funds and state capital in provincial environmental protection, air pollution control, technical transformation of industrial enterprises, energy saving and emission reduction, strategic emerging industries, scientific and technological innovation, and give priority to support expenses related to air pollution control.

(2) Ecological restoration

In 2014, Hebei invested around 70 billion yuan in projects including comprehensive management of groundwater over-exploitation pilot project, Green

Annex I. Policies on climate public expenditure

Hebei Project, the Middle Route Supporting Project for South-North Water Diversion, the Lake Filling Project for Diverting Yellow River into Hebei, the Mountain Restoration Project, Tailing Pond Comprehensive Management Project, Pollution Control for Coastal Waters in Beidaihe and Adjacent Areas, Baiyang Lake Comprehensive Management Project, Lake Wetland Protection Project, Agricultural Water-saving Irrigation Project, Heavily Polluted River Management Project, Shuangfeng Temple Reservoir Construction Project, Soil Erosion Management Project, High-standard Farmland Construction Project, Clean Agricultural Production Demonstration Construction Project, the Construction of Ecological Transition Zone of Beijing, Tianjin and Baoding, River Network Construction, Ecological Function Zone for Zhangcheng Water Conservation, Restoration of Degraded Forest in Zhangjiakou Bashang Area. In 2016, in order to protect the forest land and stop the commercial logging of natural forests, subsidies are provided for state-owned natural forest in accordance with the approved amount of forest approved by the State Forestry Administration, which equal to RMB 1,000 (US$156) per cubic meter. The management subsidy for state-owned natural forest is RMB 6 (US$1) per mu, and the management subsidy for collective and individual natural forest is RMB 15 (US$2.3) per mu annually.

(3) Groundwater over-exploitation management

Both the quantity and area of over-exploited groundwater in Hebei account for 1/3 of the total number in the country. Since 2014, the state has initiated groundwater over-exploitation comprehensive management pilot project in Hebei. Clarify water rights, develop water prices and vigorously implement the comprehensive reform of water prices; control the total amount of groundwater, strengthen management and strictly monitor the groundwater exploitation; save local water resources, divert water from other areas, focus on the development of modern water-saving agriculture and increase the alternative water sources; reduce groundwater exploitation and restore groundwater ecology via comprehensive management. In 2014, comprehensive management of groundwater over-exploitation was implemented in 49 countries of 4 cities, with a total investment of RMB 7.49 billion (US$1.17 billion). In 2015, the pilots were expanded to 63 counties of 5 cities, with a total investment of RMB 8.26 billion (US$1.29 billion). In 2016, the pilots are expanded to a total of 115 counties (cities, districts) in 9 cities and 2 provincial counties, with a total investment of RMB 8.712 billion (US$1.36 billion). The pilot projects from 2014 and 2015 had resulted in a reduction of 1.52 billion cubic meters of agricultural groundwater exploitation.

The Pilot Plan for Comprehensive Management of Groundwater Overexploitation in Hebei (2016) clearly states that the government will provide subsidies for groundwater over-exploitation management projects to guide the farmers and agriculture entities to participate in adjusting planting structure and taking water-saving measures actively. On the adjustment of planting mode projects, 500 yuan (US$ 78) per mu is provided as subsidies. For dry farming projects, a subsidy of 100 yuan (US$ 16) per Mu will be granted. For non-crop alternative projects, a subsidy of 1,500 yuan (US$ 234) per mu will be provided, which will continue for 5 years and begins to halve from the second year. For the promotion of the winter wheat water-saving technology, a material subsidy of 75 yuan (US$ 11.7) per mu will be granted to water-saving varieties. For the sprinkler irrigation, micro-irrigation, water and fertilizer integration and other efficient water-saving projects, a comprehensive subsidy of 1,500 yuan (US$ 234) per mu will be provided (including irrigation measurement facilities). For the pipeline water supply projects, a comprehensive subsidy of 870 yuan (US$ 136) per mu (including irrigation measurement facilities) will be provided. For the conservation tillage project, a subsidy of 50 yuan (US$ 7.8) per mu will be provided for the operations of machines.

(4) Energy saving and emission reduction

In 2012, the 12th FYP for Energy Saving and Emission Reduction in Hebei was issued which put forward a series of goals, such as the GDP energy consumption of 2015 should be 18% less than that of 2010. In 2013, the special funds for the prevention and control of air pollution closed down backward production capacity of RMB 490 million (US$ 76.6 million). Enterprises and projects with high energy consumption, heavy pollution, backward technologies and overcapacity were shut down, including iron and steel, cement, glass, coal, paper, printing and dyeing. The year witnessed the elimination of 8347 enterprises, reduction of 7.88 million tons of crude steel capacity, 5.86 million tons of iron, 17.16 million tons of cement, 14.88 million standard weight box of flat glass. The elimination further optimized the economic structure, and reduced the pressure on environmental governance. Vigorously promote the 6643 Project, which is to reduce 60 million tons of steel, 60 million tons of cement, 40 million tons of coal and 30 million standard weight boxes of glass by 2017. Other related projects include the new and old "Double Thirty", "Double Thousand", innovation of key practical technologies in key energy-saving industries, the "Sunday Action" to reduce the excess capacity of steel, and the

Annex I. Policies on climate public expenditure

concentrated operations to reduce the excess capacity of cement. For the special operations including changing fuel from coal to gas, shut down of clay and brick kilns, "chimney demolition" and the elimination of yellow label cars, Hebei (the provincial Department of Finance) provides priority capital incentives to places where the tasks are overfulfilled through financial budget control, which has encouraged and mobilized the enthusiasm of local governments to prevent pollution and ensured that the key annual targets of the provincial government could be achieved on time.

(5) Energy structure optimization

To improve the energy mix, while implementing centralized heating, a series of measures have been implemented including changing fuel from coal to natural gas, changing fuel from coal to electricity, promoting the use of clean coal, replacing coal-fired boiler, advocating the use of clean coal, briquette, biomass, solar power, geothermal and other clean forms of energy.

Coal burning is a major source of air pollution. Hebei invested RMB 810 million (US$ 126.6 million) in total in 2014 to eliminate in advance or upgrade coal-fired boilers for energy-saving and environmental protection. A subsidy of 20,000 yuan (US$ 3,125) per ton of steam at most will be provided to the boilers reaching the second or higher energy efficiency standards after the upgrading, and the subsidy fund for a single project cannot exceed 50% of the total investment. For the coal boilers eliminated through removal, replacement or renewal, a subsidy of 20,000 yuan (US$ 3,125) per ton of steam at most will be granted.

Hebei Province started to the implement the coal-fired boiler management in 2015, and by the end of 2017, it aims to complete the phase-out task of 11,071 sets of coal-fired boilers. For the remaining 23,562 coal-fired boilers, it is intended to ensure a comprehensive upgrading for energy saving and environmental protection with the target of 30%, 30% and 40% by 2015, 2016 and 2017 respectively.

In 2015, through upgrading to ultra-low-emission coal-fired units, shutting down solid clay brick kilns and renovating coal-fired boilers, Hebei reduced 5 million tons of coal consumption. So far 12,009 coal-fired boilers have been phased out, the capacity of which equals to nearly 3 million tons of steam. Within built-areas, coal-fired boilers with capacity of or below 10 tons of steam have all been eliminated. 252 low-emission coal-fired units have been completed. 2,780 solid clay kilns were shut down, which completely stopped the direct emissions at a low altitude from 2.6 million tons of coals.

The Special Action Plan for Coking Industry Pollution Remediation in Hebei issued in April 2016 proposed to reduce coal production capacity of 6 million tons by the end of the year through transformation to ensure all existing coke production enterprises meet pollutant discharge requirements.

The Special Action Plan for Open Mine Pollution Remediation in Hebei issued in April 2016 proposed in-depth pollution treatment of 1,881 open mines within three years. On the basis of returning in proportion the cost of mining rights and governance deposit, by the end of June 2016, the open mines applying for closure will be given appropriate compensation and incentives. For the open mines with certificates that were forced to close their business for violating the laws and regulations, the cost of mining rights and governance deposit will not be returned. Nor will they be compensated.

The Special Action Plan for Scattered Coal Pollution Remediation in Hebei issued in June 2016 proposed comprehensive treatment of the production, circulation and use of scattered coal in the province within three years.

In 2016, the Hebei provincial government promulgated the Guiding Opinions on Accelerating the Change of Fuel from Coal to Electricity and Natural Gas in the Coal-free Zone in Baoding and Langfang. In order to improve the ecological environment and quality of life, by the end of October 2017, realize zero coal burning except for thermal coal, concentrated heating and raw material coal. To spur the change of fuel from coal to electricity and natural gas in the rural areas while taking into account the economic capacity of the public, a series of subsidies will be granted for the countryside changing fuel from coal to electricity and natural gas, including equipment purchase subsidies, electricity and gas subsidies, and being able not to follow the tiered electricity and natural gas pricing. Specifically, equipment purchase subsidies for replacing coal with electricity and natural gas are 85% and 70% respectively, and the maximum amount of subsidies per household should not exceed 7,400 yuan, and 2,700 yuan respectively; during the heating period, the subsidy for residential electricity is 0.2 yuan/kWh and the maximum subsidy of electricity quantity for each household is 10,000 kilowatt hours; the natural gas subsidy is 1 yuan/cubic meter and the maximum annual subsidy of gas quantity per household is 1,200 cubic meters.

From June 1, 2015, the Decision of People's Congress Standing Committee of Hebei on the Promoting Comprehensive Utilization of Crop Straw and Banning Open Burning started to be implemented. Based on the original special funds of Hebei for banning open crop straw burning, another special funds for the comprehensive utilization of straw was set up, with a focus on grinding straw in a mechanized way and returning it to the field, turning straw silage into fodder, establishing a straw collection and storage service system, promoting biomass stove and resource utilization such as straw gasification and curing molding.

(6) Energy efficiency improvement

Promote industrial technical transformation and upgrading. In 2013, a total fund of RMB1.05billion (US$164.1million) was invested in the technical transformation of industrial enterprises to provide bank discount and special support for the key transformation and construction projects. During the year, a total of 1124 key transformation and construction projects were promoted, which guided RMB 720 billion (US$112.5 billion) of industrial investment in technical transformation and promoted the development of strategic emerging industries[①], traditional enterprises with advantages and modern service industry.

Between 2011 - 2013, a total of 594 energy saving and emission reduction technology projects from industrial enterprises in Hebei were supported by the national and provincial special funds, including 388 energy-saving emission reduction technical transformation projects for energy saving and emission reduction, 39 key energy management center for industry and enterprises in Hebei, 43 clean production technology demonstration projects, 115 technical transformation projects for comprehensive utilization of resources, 58 technical transformation projects for energy-saving and environmental protection products. With a total of RMB 2.33 billion (US$364.1 million) support from national and provincial funding, the implementation of these projects effectively alleviated the growing trend of major pollutants industrial emissions of major pollutants, and improved the quality of the environment.

In 2013, the Department of Finance in Hebei raised RMB 56 million (US$8.75 million) of funds in total, with a focus on supporting 28 enterprises projects to build capacity for innovation. In order to increase the support for the strategic emerging industries, Hebei set up a special fund for strategic emerging industries starting from 2012. The province would provide 1 billion yuan each year as a special fund to support the strategic emerging industries, including key technology research and

---

① The strategic emerging industries refer to the seven industries of great significance that the central government of China put forward in 2010 to spur the upgrading of industrial structure and transformation of economic development mode, enhance China's independent development capacity and international competitiveness, and promote economic and social sustainable development, including energy efficient and environmental technologies, next generation information technology (IT), biotechnology, high-end equipment manufacturing, new energy, new materials, and new-energy vehicles (NEVs).

development, industrialization of high-tech achievements, enhancing innovative capacity, key application demonstration, cultivation of high-growth enterprises, introduction of the large leading projects, industrial innovation and development and regional agglomeration development. In 2013, the provincial budget provided 1.6 billion yuan of special funds for central enterprises to move into Hebei, strategic emerging industries, and development of modern logistics industry. The provincial investment in 2014 reached 4.38 billion yuan of special funds with a focus on the main industrial development. Through financial policy to support the middle and central economic zone of Hebei to develop advanced manufacturing and specific industries and strive to establish an industrial structure with a clear division of labor, complementary advantages, distinctive features as well as coordinated development.

Through the special funds for strategic emerging industries, the special funds for the development of modern logistics industry, special funds for the central enterprises to move into Hebei, Hebei focuses on industries with great market potential, solid industrial base and leading roles, accelerates the formation of pillar industries, highlights technological innovation and emerging industries development and spurs the transformation of economic development mode.

In 2014, the research and development funds that Hebei invested in the prevention and control of air pollution as well as energy saving and emission reduction reached over RMB 50 million (US$ 7.8 million). Furthermore, based on the "135" Project, Hebei has also optimized the allocation of scientific and technological resources inside and outside the province, increased investment in science and technology year by year, implemented a number of major science and technology projects in line with the priorities, and constantly broken through the bottleneck of air pollution control technologies, and provided comprehensive technical support for coal burning reduction, vehicle and oil control, pollution and emission management, and air cleaning as well as dust reduction.

(7) Comprehensive agricultural development

In recent years, the annual financial investment in Hebei to support food production has continued to increase by almost 30%. In 2013, Hebei issued RMB 6.83 billion (US$ 1.07 billion) of grain subsidies; invested RMB 1.738 billion (US$ 278.59 million) of funds to support infrastructure construction for farmland water conservancy, which solved the watering problem of 19.09 million mu of farmland; invested RMB 447 million (US$ 69.8 million) as agricultural relief funds for disaster rescue and relief, restoration of damaged water conservancy projects, agricultural production relief and facilities reconstruction; granted RMB 1.26 billion (US$ 196.9 million) of funds for the fine breeds of wheat, corn, rice, cotton and other crops; provided RMB 1 billion (US $156.3 million) of subsidies for agricultural machinery purchase; RMB 265 million (US$ 41.4 million) of special incentives was issued to 77 advanced grain production counties and cities with the total grain output increasing by over 3.4% in 2011 in order to support grain production. 443 million yuan of provincial subsidy funds was allocated to support the land management projects for comprehensive agricultural development by the central government. RMB 416 million (US$ 65 million) of provincial subsidy funds for the key county projects of small agricultural water conservancy was granted to support the central government; RMB 250 million (US$ 39.1 million) of provincial funds was arranged to carry out sub-soiling operations for 10 million mu of farmland in the province; RMB 163 million (US$ 25.5 million) of county and provincial subsidy funds was granted for modern agricultural projects in the province to support the modern agricultural production and development projects by the central government; RMB 110 million (US$ 17.2 million) of central subsidy funds was raised to build high yield agriculture and establish 10,000 mu of demonstration farmland with wheat, corn, soybeans, potatoes and grain.

In 2014, RMB 6.83 billion (US$ 1.07 billion) of benefitting-farmer financial subsidies was allocated to cities and counties (of which grain subsidies took up RMB 780 million (US $121.9 million) and agricultural subsidies RMB 6.05 billion (US$ 0.95 billion)), which covered a total area of 78.9 million mu.

In 2015, the central government's comprehensive agricultural development funds reached RMB 2.024 billion (US$ 316.25 million). At the same time, loans, discounts, financial subsidies and other means have been adopted to attract social funds actively. The total provincial investment in agricultural development reached RMB 2.748 billion (US$ 429.375 million). RMB 50 million (US$ 7.8 million) of provincial special funds was arranged to transform, upgrade and construct 100 provincial modern vegetable industry parks. Since January 1, 2016, the proportion of agricultural insurance subsidies for three major food crops, i.e. rice, wheat and corn in the major grain production counties was adjusted: the proportion of central and provincial financial subsidies were increased; the proportion of municipal financial subsidies remained unchanged; the proportion of county financial subsidies were reduced to zero. Since July 1, 2016, Hebei has started pilot insurance subsidies for featured agricultural industries and products.

(8) Rural transformation

To promote the integration of urban and rural areas and the equalization of public services, in 2014, Hebei proposed to build 3,000 beautiful villages and fully complete the 15 practical tasks for the transformation and upgrading. Hebei allocated 143 million yuan as provincial subsidy funds for the planning of the main villages, garbage removal and energy-saving renovation of residential houses in pilot villages. Energy-saving special incentive funds were granted to 4,417 households from the 10 pilot villages. The standard of the incentives was 5,000 yuan per household and the total amount was RMB 22.085 million (US$ 3.45 million).

In 2014, Hebei was identified as a key province for the special fund of the Food Security Project by the central government in order to repair local dangerous and outdated warehouses. Hebei plans to invest RMB 1.2 billion (US$ 187.5 million) (of which the central subsidies account for 300 million yuan) to conduct retrofitting and reconstruction of the state-owned grain warehouses in 11 cities and 151 counties (cities, districts).

Since 2015, Hebei has strengthened the support for retrofitting dilapidated houses in poor rural areas. The retrofitting fund for dilapidated rural houses gives priority to the counties where 7366 poor villages are located, while also taking into account the key villages in other counties (cities or districts) working on rural landscape transformation as well as the areas with high-intensity earthquake fortification. As of September 2015, Hebei has allocated RMB 1.6 billion (US$ 250 million) of retrofitting funds for dilapidated rural houses, of which the central capital accounted for RMB 1.045 billion (US$ 163.3 million) and the provincial matching funds were 558 million yuan. The dilapidated houses of 123,000 households have been retrofitted, of which 30,000 are set as energy-saving model households.

(9) Flood prevention and relief

In order to better conduct the flood prevention and relief work and alleviate the impact of "7.19" flood, in 2016, Hebei allocated a total of RMB 2.587 billion from the central and provincial disaster relief funds, which provided a strong financial guarantee for the disaster relief and post-disaster reconstruction and recovery in the disaster-stricken area.

2. Innovative expenditure patterns

(1) Public and private partnership

On April 30, 2015, Hebei released the first batch of project list to encourage private investment in traffic, energy, municipal services and other areas. It involved a total of 38 projects on expressway, first-grade highway, railway, clean energy, cogeneration, hydropower generation, military production, medical facilities, as well as urban water supply, heating and sewage treatment, with a total investment of RMB 210. 61 billion (US$ 32. 9 billion) to encourage private capital to participate in project construction and operation through joint ventures, sole proprietorship, equity participation, franchise, and etc. In May 2016, the Hebei Provincial Price Bureau and the Provincial Department of Housing and Urban-Rural Development issued the *Opinions on the Implementation of Paid Use System for the Urban Underground Integrated Corridor*, requiring all cities to establish and improve the paid use system for the urban underground integrated corridor that will both attract social capital to participate in corridor construction, operation and management, and contribute to mobilizing the enthusiasm of the units along the pipeline to join the corridor construction. The Opinion came into effect on June 1, 2016 and will be valid for 5 years. By the end of 2016, Hebei has 529 projects in reserve with a total investment of more than 1 trillion yuan, among which 42 projects were implemented with an investment of RMB 163. 3 billion (US$ 25. 5 billion).

(2) International loans

On December 10, 2015, the Executive Board of the Asian Development Bank (ADB) approved the air pollution control project in Hebei and would provide policy loans worth US$ 300 million for the Beijing-Tianjin-Hebei Air Quality Improvement – Hebei Policy Reforms Program. This is the first time that ADB has granted policy loans to China. Together with the € 150 million euros of joint financing to be provided by the KFW Bankengruppe, the funds will jointly promote the air pollution control in Beijing-Tianjin-Hebei. On June 6, 2016, the Executive Board of the World Bank formally approved the China – Hebei Air Pollution Prevention and Control Program, the loan amount of which is US$ 500 million with a maturity of 19 years. It is the first World Bank results-oriented loan program in China. After the loan is disbursed, an equity investment fund will be set up together with the US$ 300 million policy loan from ADB and the €150 million from the KFW Bankengruppe to give full play to the guiding and stimulating role of capital and to leverage more social capital to the air pollution prevention and control projects as well as related industries in Hebei.

(3) Clean Development Mechanism Fund

The State Council approved the establishment of the China Clean Development

Mechanism Fund (CDM Fund) in 2007. Since 2011, Hebei, as one of the earliest province to use the loan, has started the compensatory use of the CDM Fund. By the end of 2014, Hebei has used RMB 595 million (US$ 93 million) of preferential loans from the CDM Fund.

(4) Emissions trading

In 2013, the Hebei Department of Finance joined hands with the Department of Environmental Protection to vigorously promote emissions trading, actively explore policies on sewage charges for volatile organic compounds and environmental insurance for heavily polluting corporate, and provide guiding funds to promote cleaner production demonstration in the major industries. With these policy supports, enterprises get not only financial support but loan support from financial institutions. In 2013, the 13 enterprises who acquired compensatory emissions permits in the province got RMB 34.448 million (US$ 5.38 million) of emission pledge loans provided by the Shijiazhuang Branch of Everbright Bank.

In December 2014, the construction of Beijing-Hebei cross-regional carbon emissions trading market started. The market will establish a cross-regional unified accounting method, verification standard and trading platform to explore the implementation of the carbon emissions trading system and establish a unified carbon market in the country. The cross-regional carbon emissions trading market adopts the quota trading mechanism under the control of the total amount of carbon dioxide emissions, and the trading products include carbon emission quotas and certified emission reductions (emission reduction from Certified Emission Reduction, energy-saving projects and forestry carbon sequestration projects). The trade subjects in the two cities can buy and sell emission quotas and certified emission reductions freely which can be used for compliance.

Annex II. Related government bodies

Government bodies related to climate change in Hebei

NO.	Related government bodies	Main responsibilities in tackling climate change	Related divisions
1	Hebei Provincial Development and Reform Commission	- Develop planning and policy measures for circular economy, and coordinate the implementation; participate in the preparation of environmental protection planning, coordinate the promotion of environmental protection industries and cleaner production; lead the project management of energy and resource conservation, renewable resources, enterprise and urban sewage treatment and urban waste disposal; organize the formulation of the province's strategies, planning and related policies for tackling climate change, and organize the implementation of the clean development mechanism; undertake the daily work of the office for the leading work team on tackling climate change, energy conservation and emission reduction; guide the promotion of bulk cement and energy conservation supervision; - Organize the development of agriculture and rural areas, ecological environment construction; - Coordinate the convergence between the energy & transportation development planning and the national economic and social development planning; - Conduct comprehensive analysis of the coordinated development of the province's economy & society and resources & environment; formulate policies on the conservation and comprehensive utilization of energy and resources, and put forward strategic recommendations for energy development; major issues in energy development and reform; participate in the preparation of suggestion for the total energy consumption control, guide and supervise related work on the total energy consumption control; responsible for the statistical analysis of energy industry and forecast as well as warning. - formulate development planning and policies of coal development, coalbed methane and processing coal into clean energy products and organize the implementation; - develop plans, annual plans and policies of thermal power, nuclear power and power grid (excluding rural power grids) and organize the implementation; - guide and coordinate the development of new energy, renewable energy and rural energy; - guide energy conservation and comprehensive utilization of resources in the energy industry; industrial management for oil, natural gas, oil refining, coal natural gas, coal fuel and biomass liquid fuel; - organize the formulation of the development and setup planning for the energy market; regulate energy market operation; deal with energy market disputes; propose suggestion on the adjustment of energy price;	Tackling climate change Division, Rural Economy Division, Basic Industry Division, General Section of the Energy Bureau, Coal Division of the Energy Bureau, New Energy Division of the Energy Bureau, Energy Market Supervision Division of the Energy Bureau.

Con.

NO.	Related government bodies	Main responsibilities in tackling climate change	Related divisions
2	Hebei Provincial Department of Finance	- Comprehensively manage national fiscal revenue and expenditure, fiscal and taxation policies, implement financial supervision and participate in the macro-control of national economy. - Establish management system for the loans (donation) from international financial organizations, foreign governments and clean development trust; responsible for the application and lending of loan (donation) from international financial organizations, foreign governments and clean development trust as well as fund use, repayment and statistics; undertake foreign economic cooperation and exchanges.	Taxation Division, Social Security Division, Resource and Environment Division, Economic Development Division, Budget Division, Agriculture Bureau, General Division, PPP Office, Procurement Office
3	Hebei Provincial Department of Environmental Protection	- Establish and improve the core system of environmental protection. Formulate and organize the implementation of the province's environmental protection policies and planning, and draft local laws and regulations. Develop and supervise the implementation of pollution prevention planning in the major areas and watershed, as well the environmental protection planning for drinking water source. In line with the requirements of the provincial government, join hands with the relevant departments to develop pollution control planning for key sea areas and participate in the development of the main functional zoning in the province; - Coordinate and supervise major environmental issues. Take the lead in coordinating the investigation and handling of major environmental pollution and ecological damage accidents in the province, guide and coordinate the municipal and county governments in the emergency response and early warning for major environmental accidents, coordinate and resolve the disputes concerning cross-regional environmental pollution, coordinate the pollution prevention and control of key watersheds, regions and waters, and guide, coordinate and supervise marine environmental protection work; - Undertake the responsibility to implement the province's emission reduction targets. Establish and oversee the implementation of the mechanism for major pollutant discharge control and emission permits; - Propose opinions on the scale and direction of fixed assets investment in the field of environmental protection as well as on provincial fiscal fund arrangement, participate in guiding and promoting the development of circular economy and environmental protection industry in the province, and participate in tackling climate change; - Assume responsibility for the prevention and control of environmental pollution and damage from the source. Commissioned	Vehicle Air Pollution Control Division, Air Pollution Prevention and Control Division, Water Pollution Control Division, Rural Environmental Protection Division, Environmental Monitoring and Emergency Response Division, Radiation Safety Management Division (Nuclear

Annex II. Related government bodies

NO.	Related government bodies	Main responsibilities in tackling climate change	Related divisions
3	Hebei Provincial Department of Environmental Protection	by the provincial government, conduct environmental impact assessment on major economic and technological policies, development planning as well as major economic development plans; provide opinions on local laws and regulations and draft concerning environmental protection; review and approve the environmental impact assessment documents of major development and construction area and projects according to the state and provincial regulations; – Supervise and manage environmental pollution prevention and control. Formulate the pollution prevention and control management system for water, atmosphere, soil, noise, light, malodor, solid waste, chemicals and motor vehicles and implement it; supervise and manage the environmental protection work for drinking water sources in conjunction with the relevant departments; and organize and guide urban and rural comprehensive environmental remediation work; – Guide, coordinate and supervise the ecological protection work. Formulate ecological protection planning, organize quality assessment for ecological environment, and supervise the development and utilization of natural resources which have effect on the ecological environment, the construction of key ecological environment and ecological restoration work. Guide, coordinate and supervise the environmental protection work of various types of nature reserves, scenic spots and forest parks; coordinate and supervise the protection of wild animals and plants, wetland environmental protection and desertification control work. Coordinate and guide the protection of rural ecological environment; supervise the environmental safety of biotechnology; lead the work of biological species (including genetic resources); organize and coordinate biodiversity conservation; – Responsible for environmental monitoring and information dissemination. Develop environmental monitoring mechanism and regulations; organize and implement the monitoring of environmental quality and pollution sources. Organize investigation and evaluation of the environmental quality and issue forecast and early warning; organize the establishment and management of provincial environmental monitoring network and information network; establish and implement environmental quality notice system; release the province's comprehensive environmental report and major environmental information; – conduct environmental protection science and technology work; organize scientific research and technical engineering demonstration on environmental protection; promote the construction of environmental technology management system; – conduct international cooperation and exchanges on environmental protection; put forward suggestions for relevant issues in international and inter-provincial environmental cooperation; organize and coordinate the implementation of international treaties on environmental protection; and participate in the handling of foreign-related environmental protection affairs; – Organize, guide and coordinate environmental education and publicity work; formulate and organize the implementation of environmental protection publicity and education program; carry out publicity and education work on ecological civilization construction and environment-friendly society construction; and promote the public and social organizations to participate in environmental protection.	Safety Management Division), Total Pollutant Discharge Control Division, Policy and Regulation Division

NO.	Related government bodies	Main responsibilities in tackling climate change	Related divisions
4	Hebei Provincial Department of Industry and Information Technology	The division is mainly responsible for the formulation and implementation of the province's industrial energy conservation and comprehensive utilization of resources and cleaner production promotion policies; participating in the development of planning for energy conservation and comprehensive utilization of resources, cleaner production promotion and pollution control policies; proposing opinions on the energy and water consumption of the industrial sector that needs approval of the provincial government; organize and guide the industrial energy-saving equipment (product) manufacturing and enterprises' energy management; organize and coordinate major demonstration projects and the promotion of new products, new technologies, new equipment, and new materials; study, formulate and implement policies on the comprehensive utilization of resources and project management of industrial "three wastes".	Energy Conservation and Comprehensive Utilization Division
5	Hebei Provincial Department of Forestry	- Organize, coordinate, guide and supervise the afforestation work in the province. Develop guiding plan for afforestation in the province to guide the control of soil erosion through biological measures including afforestation, forest conservation and planting trees and grass; guide and supervise the public to voluntarily plant trees and support afforestation work. Undertake forestry related work on tackling climate change; - Undertake the responsibility to supervise and manage the protection and development of forest resources in the province; - Organize, coordinate, guide and supervise the wetland protection work in the province. Formulate provincial wetland protection planning as well as local standards and regulations for wetland protection; organize and implement the protection and management work including the construction of wetland protection zones and wetland parks; and supervise rational use of wetlands; - Organize, coordinate, guide and supervise the desertification control work in the province. Organize the formulation of planning for the provincial prevention and control of desertification and stony desertification as well as desertified land conservation reverses construction; supervise the rational use of desertified land; organize and guide the review of project impact on land desertification; and organize and guide the forecast, early warning and emergency response of sandstorm; - Supervise and manage the nature reserves of forestry system in the province. Under the guidance of national and provincial nature reserve zoning and planning principles, guide legally the construction and management of nature reserves in the form of forest, wetland, desertification and terrestrial wildlife; supervise and manage forestry biological germplasm resources, new plant varieties and biodiversity conservation; - guide and supervise the development and utilization of forest, wetland, desert and terrestrial wild animals and plant resources by various industries across the province; - Participate in the formulation of economic regulation policies for forestry and its ecological construction; organize and guide the establishment and implementation of ecological compensation system for forestry and its ecological construction. Prepare departmental budgets and organize the implementation; supervise state-owned forestry assets and forest resources assets; manage provincial forestry funds; guide and supervise the management and use of provincial forestry funds.	Afforestation Management Division, Forest Resources Management and Policy Regulations Division, Development Planning and Fund Management Division

Annex II. Related government bodies

NO.	Related government bodies	Main responsibilities in tackling climate change	Related divisions
Con.			
6	Hebei Provincial Department of Agriculture	- Responsible for the protection of agricultural resources. Guide the protection, management and efficient use of agricultural land, fishery waters, grassland, agriculture-suitable tidal flat, agriculture-suitable wetlands and agricultural biological species resources; - Develop and implement construction planning for eco-agriculture and renewable energy. Guide the comprehensive development and utilization of agricultural resources and rural renewable energy, agricultural biomass industry development, and rural energy saving and emission reduction; undertake agricultural non-point source pollution control; designate areas where agricultural production is prohibited; guide the development of ecological agriculture and circular agriculture; protect the ecological environment of fishery waters.	Beautiful Village Construction Division, Modern Agriculture Park Division (Mountainous Area Comprehensive Development Office), Agricultural Resources and Environment Division
7	Hebei Provincial Meteorological Bureau	- In the administrative area, organize cross-regional and cross-sectoral joint monitoring and forecast for major disastrous weather, timely propose meteorological disaster prevention measures, and conduct assessment on major meteorological disasters with the aim to provide decision-making basis for the government to prepare against the meteorological disasters; manage the public meteorological services in the administrative area; manage the release of professional meteorological forecast in the administrative area including public weather forecast, disastrous weather warning, agricultural weather forecast, meteorological forecast of the urban environment and meteorological forecast of the fire danger rating. - Propose suggestions to the people's governments and relevant departments at the same level to utilize and protect climate resources and to promote climate resources zoning and other achievements; organize climate feasibility studies for climate resources development and utilization projects; participate in the tackling climate change work by the provincial government; organize and conduct climate change impact assessment, technology development and decision-making advisory services.	

NO.	Related government bodies	Main responsibilities in tackling climate change	Related divisions
8	Hebei Provincial Department of Housing and Urban-Rural Development	- Develop the long-term planning, reform measures, rules and regulations and technical standards for urban construction and municipal public utilities; guide the urban water supply, water saving, gas, heat, municipal facilities, gardens, city environment management, urban construction monitoring, water security supervision of urban landscape and other work; guide the construction of urban sewage treatment facilities and supporting pipe network; guide the construction of harmless disposal of urban garbage and supporting facilities; guide the greening of urban planning areas; review, approve, supervise and manage the scenic spots above provincial level. - Formulate policies and development plans for the construction machinery and equipment industry and supervise the implementation; guide the renovation of wall materials; - Guide and standardize the construction machinery and equipment market; regularly issue catalog of construction machinery and equipment that are under elimination, restriction or promotion; guide and manage the use of energy-saving construction materials and products.	Urban Construction Division, Construction Material and Equipment Division
9	Hebei Provincial Department of Water Resources	- In accordance with the relevant laws, regulations and standards of national resources and environmental protection, develop water resources protection planning; organize water functional area zoning and water discharge control of different functional areas; monitor the water quantity and quality in rivers, lakes and reservoirs, examine the assimilative capacity of the watershed, and propose opinions on the limit of total discharge volume. - Develop economic regulation measures for the province's water conservancy industry; conduct macro-adjustment of the use of water resource fund; guide the water supply industry, hydropower and other businesses; implement national policies on the assets, pricing, tax, credit and finance of water conservancy, and coordinate with relevant departments to develop the province's policy measures and organize the implementation; in line with relevant national provisions, supervise and manage state-owned assets in the water conservancy system. - Organize the implementation of water resources abstraction permit, water resources paid use system and water resources demonstration system; organize water resources survey, evaluation and monitoring; guide and supervise the implementation of water allocation, water function zoning and water resources dispatch; organize the formulation of water resources protection planning; guide the protection of drinking water sources, urban water supply planning, urban flood prevention, urban sewage treatment and other non-traditional water resources development work; guide the setup of the sewage outlets to the river; guide the planned water use and water conservation work.	Water Resources Division, Water Protection Division

Source: Consolidated data from the website of Hebei Provincial Government and field survey.

Annex III. The main contents of budget management reform in Hebei

1. Improve government budget system

(1) Implement overall budget management. Include all the revenue and expenditure of the provincial government in the budget management. Based on a clear scope of revenue and expenditure, prepare general public budget①, government fund budget②, state capital operating budget③ and social insurance fund budget④, and

---

① General public budget is the fiscal revenue with tax revenue as the main part, which is used to protect and improve of people's livelihood, promote economic and social development, safeguard national security, maintain the normal operation of national institutions and others.

② The government fund budget is the fund levied, collected or raised in other ways from specific objects within a certain period of time in accordance with the provisions of laws and administrative regulations, which is specially used for budget for revenues and expenditures for the development of specific public utilities. From January 1, 2015, among the government fund budget, the project revenue and expenditure for the provision of basic public services and the operation of personnel and institutions will be integrated into the, general public budget including local education surtax, cultural undertakings construction fees, employment insurance fund for the disabled, funds from the provisions of local land transfer for farmland and water conservancy construction as well as education, income from transfer of road toll right for government loan repayment, afforestation fund, expenditure for forest vegetation restoration, the water conservancy construction fund, harbor dues, and channel maintenance fee for Yangtze estuary.

③ The state capital operating budget is the revenue and expenditure from the state capital gains. The state capital operating budget is prepared in accordance with the principle of balance of payments without deficit, and transfers funds to the general public budget. In addition to transferring fund to the general public budget and supplementary social security fund, the budget is limited to solving the historical problems and reform costs of state-owned enterprises, capital injection for state-owned enterprises and policy subsidies. The funds of general public budget for these aspects will withdrawal gradually.

④ Social insurance fund budget is the fund from social insurance contributions, general public budget arrangements and fund collected from other means, which is specially devoted to the revenues and expenditures of social insurance. Social insurance fund budget is prepared in accordance with the overall level and social insurance projects, and should ensure balance of payments.

establish a government budget system with clear positioning and division of labor. Government fund budget, state capital operating budget, social insurance fund budget are linked with general public budget.

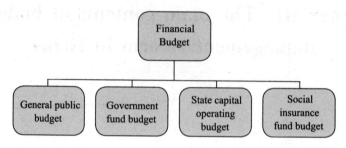

Figure 1

The four types of budget have different focuses, and fiscal expenditure on tackling climate change is included in the general public budget. Thus, the following analysis primarily targets relevant expenditures in the general public budget.

(2) Strengthen coordination among government funds, state capital operating budget and general public budget. In accordance with the central deployment, gradually abolish the regulations of special funds for special purpose, including urban maintenance and construction tax, sewage charges, cost of prospecting and mining rights, and compensation for mineral resources. The relevant funds will be arranged from the general public budget. Allocate the budget in a unified way, and gradually include all the budget funds into the financial sector for overall arrangement.

(3) Strengthen the overall consideration and arrangement of funds between the corresponding level and the next higher level. Hebei provincial government prepares the budget at the beginning of the year, includes transfer payments at higher level such as tax returns at the higher level, revenue collected from the lower level, those listed in the base and noticed in advance, and the revenue of the corresponding level into the revenue budget, and allocate the expenditure of the corresponding level and payment to lower level in a manner of overall consideration and arrangement.

(4) Improve the budget standard system. Hebei Province gives full play to the basic supporting role of expenditure standards in budget preparation and management, perfects the quota standard system for basic expenditures, improves the quota and service standards for institutional operating funds, and speed up the establishment of the quota standard system for project expenditure. Establish a dynamic adjustment mechanism for the quota standards, and make appropriate adjustment based on economic and social development as well as policy changes. Strengthen the staffing and asset

Annex III. The main contents of budget management reform in Hebei

management and improve the mechanism that integrates staffing management, asset management and budget preparation. Establish information base for departmental budget, and enhance the budget management foundation for basic expenditure.

2. Improve budget control methods

(1) Adopt medium-term financial planning and management

Starting from the preparation of the 2016 budget, Hebei Province has been formulating the mid-term financial planning. According to the economic operation as well as the direction of macro-control, provide scientific forecast of the fiscal revenue in the next three years, make a comprehensive combing analysis of the major reforms and expenditure policies, prepare a 3-year rolling financial planning of the corresponding level in a manner of overall consideration and arrangement, and link up with the local economic & social development plan and national macro-control policies. The preparation of annual budget is linked to the medium-term financial plan. A department at any level, when formulating departmental and industrial planning that involve policy and financial support, should link with and the medium-term financial planning. Strengthen the construction of the budget item base, improve the project reporting and audit mechanism, and achieve the rolling management of budget items.

(2) Establish multiyear budget balance mechanism

Hebei Province has set budget stability adjustment funds for the general public budget at any level to make up for the deficiency of future budget funds. If there is any excess revenue while implementing the general public budget, it will be used for government debt or supplementary stability adjustment funds; if there is revenue deficiency, realize balance through transferring the budget stability adjustment funds or other budget funds or reducing expenditure. If balance still could not be achieved with the above measures, increase the deficit at the provincial level with the approval by the Provincial People's Congress or its Standing Committee, and report to the Ministry of Finance for the record, in the next year to make up for the budget; cities and counties can achieve balance by applying for temporary assistance from government at higher levels, and return the funds in the next year's budget. Excess revenue in government fund budget or state capital operating budget will be carried forward to the next year; if there is revenue deficiency, achieve balance by reducing expenditure.

3. Deepen performance budget reform

(1) Comprehensively promote performance budget reform

Opinions of Hebei Provincial People's Government on Deepening the Performance Budget Management Reform (Ji Zheng [2014] NO. 76) requires to speed up the establishment of a new mechanism of whole-process performance budget management. In 2015, fully implemented at the corresponding levels of 11 cities divided into districts and Dingzhou as well as Xinji, and three counties (cities or districts) were selected for simultaneous pilot project; in 2016, the policy was implemented in all cities and counties (cities or districts) in the province.

(2) Improve budget review methods

When financial departments at any level in Hebei review budget, firstly, they review whether departmental responsibilities and objectives match with government work; secondly, review whether the indicators for the performance objectives of all the work are scientific; then review the relevance between the budget items and the responsible activities as well as the necessity of the project; finally define a reasonable budget limit to ensure the overall implementation of major government decision-making arrangements and enhance government management efficiency.

(3) Fully promote performance evaluation

Hebei Province adopts the integrated method of self-assessment and financial evaluation to conduct comprehensive performance evaluation. Departments at any level are responsible for conducting performance evaluation of budget items and comprehensive self-evaluation of the annual completion; the financial departments are responsible for conducting performance evaluation of work activities and re-evaluating the key areas and major projects. According to the budget management needs, expand the scope of performance evaluation, innovate performance evaluation methods, and extend the priorities of performance evaluation from project expenditure to departmental expenditure, policies, systems and management.

(4) Strengthen the application of evaluation results

Hebei Province has established a mechanism to link budget performance with budget arrangements, and has taken the performance evaluation results as an important evidence for adjusting expenditure structure and improving the fiscal policies as well as scientific budget arrangements. Improve the performance evaluation reporting system and performance accountability system, and enhance the disclosure of performance information.

Annex III. The main contents of budget management reform in Hebei

4. Improve financial input methods

(1) Increase government purchase of services provided by social forces

Hand public services that are suitable for market-based approach and can be undertaken by social forces to qualified social forces in accordance with certain methods and procedures, and governments will pay the cost based on the quantity and quality of services. For all management services of government affairs that are suitable to purchase from social forces should, in principle, introduce a competition mechanism, purchase through contract or commission, and incorporate into the scope of administration of services purchased by government from social forces.

(2) Vigorously promote the public and private partnership model (PPP model)

Hebei Province encourages social capital to participate in the investment and operation of public services with certain benefits such as urban infrastructure through franchising, etc. For construction projects with quasi-public nature that have relatively flexible price adjustment mechanism, high degree of marketization, large scale of investment as well as long-term and stabile needs, explore the application of PPP model and leverage social capital to participate in supplying public goods.

(3) Implement subsidies after evaluation

For the science and technology development and service projects that governments support and encourage, adjust from the original subsidies in advance to the method that the institutions invest first, after gaining achievement or service performance, the financial and relevant departments review or evaluate the performance, and then provide subsidies in order to utilize the guiding role of financial funds.

(4) Actively promote equity investment

Manage various types of financial resources that governments provide to support industrial development in a manner of overall consideration and arrangement; establish equity investment funds to guide the industries; take the market approach to attract social capital to support economic and industrial restructuring and upgrading; form a capital investment mechanism that integrates fiscal and financial means.

Annex IV. Correlation between the classification of general public budget expenditure items and climate (detailed)

Item No.	Item	High Correlation	Moderate Correlation	Low Correlation
20104	Development and reform affairs		√	
2010401	Administrative operation		√	
2010402	General administrative affairs		√	
2010404	Strategic planning and implementation		√	
2010408	Price management		√	
2010450	Cause operation		√	
2010499	Other development and reform expenditure		√	
20105	Statistical information affairs			√
2010501	Administrative operation			√
2010505	Special statistical business			√
2010506	Statistical management			√
2010507	Special census activities			√
2010508	Statistical sampling survey			√
2010550	Cause operation			√
20106	Financial affairs			√
2010601	Administrative operation			√
2010602	General administrative affairs			√
2010603	Organ service			√
2010604	Budget reform business			√
2010605	Financial treasury business			√
2010606	Financial supervision			√

Annex IV. Correlation between the classification of general public budget expenditure items and climate (detailed)

Con.

Item No.	Item	High Correlation	Moderate Correlation	Low Correlation
2010650	Cause operation			√
2010699	Other fiscal expenditures			√
2040299	Other public security expenditures			√
20601	Science and technology management affairs			√
2060101	Administrative operation			√
2060102	General administrative affairs			√
2060103	Organ service			√
2060199	Other expenditures on science and technology management			√
20602	Basic research		√	
2060201	Institutional operation		√	
2060203	Natural science fund		√	
2060204	Key laboratories and related facilities		√	
2060206	Special basic research		√	
2060299	Other basic research expenditures		√	
20603	Applied research			√
2060301	Institutional operation			√
2060302	Social welfare research			√
2060399	Other applied research expenditures			√
20604	Technical research and development			√
2060401	Institutional operation			√
2060402	Application technology research and development			√
2060403	Industrial technology research and development			√
2060404	Transformation and diffusion of scientific and technological achievements			√
2060499	Other technical research and development expenditure			√
20605	Technical conditions and services			√
2060501	Institutional operation			√
2060503	Technical conditions special projects			√
2060599	Other technical conditions and service expenditures			√
20606	Social science			√

Con.

Item No.	Item	High Correlation	Moderate Correlation	Low Correlation
2060601	Social science research institutions			√
2060602	Social science research			√
20607	Popularization of science and technology			√
2060701	Institutional operation			√
2060702	Science popularization activities			√
2060703	Youth science and technology activities			√
2060704	Academic exchange activities			√
2060705	Science and technology museum			√
2060799	Other science and technology popularization expenditures			√
20608	Science and technology exchange and cooperation			√
2060801	International exchange and cooperation			√
2060899	Other expenditures on science and technology exchange and cooperation			√
20699	Other science and technology expenditures			√
2069901	Science and technology awards			√
2069999	Other science and technology expenditures			√
21101	Environmental protection management affairs		√	
2110101	Administrative operation		√	
2110102	General administrative affairs		√	
2110103	Organ service		√	
2110104	Environmental protection publicity		√	
2110199	Other environmental protection management expenditures		√	
21102	Environmental monitoring and supervision	√		
2110203	Construction Project EIA Review and Supervision	√		
2110204	Nuclear and Radiation Safety Supervision	√		
21103	Pollution prevention and control	√		
2110301	Atmosphere	√		
2110304	Solid waste and chemicals	√		
2110305	Radioactive sources and radioactive waste supervision	√		

Annex IV. Correlation between the classification of general public budget expenditure items and climate (detailed)

Con.

Item No.	Item	High Correlation	Moderate Correlation	Low Correlation
2110307	Expenditure on sewage charges	√		
2110399	Other pollution prevention and control expenditures	√		
21104	Natural ecological protection	√		
2110401	Ecological protection	√		
2110402	Rural environmental protection	√		
2110403	Nature reserve	√		
21110	Energy saving and utilization	√		
2111001	Energy saving and utilization	√		
21111	Pollution reduction	√		
2111101	Environmental monitoring and information	√		
2111102	Environmental law enforcement supervision	√		
212	Urban and rural communities expenditure			√
21201	Urban and rural communities management affairs			√
2120101	Administrative operation			√
2120105	Formulation and supervision of engineering construction standards			√
2120106	Engineering construction management			√
2120107	Municipal public sector market supervision			√
2120108	Planning and protection of key national scenic spots		√	
2120109	Housing Construction and Real Estate Market Supervision			√
2120110	Professional qualification registration and review			√
2120199	Other expenditures on urban and rural communities management			√
21202	Planning and management of urban and rural communities		√	
2120201	Planning and management of urban and rural communities		√	
21203	Public facilities of urban and rural communities		√	
2120399	Other public facilities expenditures of urban and rural communities		√	

Con.

Item No.	Item	High Correlation	Moderate Correlation	Low Correlation
21205	Environmental health of urban and rural communities		√	
2120501	Environmental health of urban and rural communities		√	
21206	Construction market management and supervision			√
2120601	Construction market management and supervision			√
213	Agriculture, forestry and water resources expenditure			√
21301	Agriculture			√
2130101	Administrative operation			√
2130102	General administrative affairs			√
2130103	Organ service			√
2130104	Cause operation			√
2130106	Science and technology transformation and promotion services			√
2130108	Pest control			√
2130109	Quality safety of agricultural products			√
2130110	Law enforcement supervision			√
2130111	Statistical monitoring and information services			√
2130112	Agricultural industry management			√
2130119	Disaster prevention and relief	√		
2130122	Agricultural means of production and technical subsidies			√
2130123	Agricultural production insurance subsidy			√
2130124	Institutionalized and industrialized agriculture management			√
2130125	Processing and marketing of agricultural products			√
2130126	Rural public welfare undertakings			√
2130135	Protection, restoration and utilization of agricultural resources	√		

Annex IV. Correlation between the classification of general public budget expenditure items and climate (detailed)

Con.

Item No.	Item	High Correlation	Moderate Correlation	Low Correlation
2130152	Subsidies for college graduates to work at the grassroots			√
2130153	Expenditure on grassland vegetation restoration			√
2130199	Other agricultural expenditures			√
21302	Forestry	√		
2130201	Administrative operation	√		
2130203	Organ service	√		
2130204	Forestry institutions	√		
2130205	Forest cultivation	√		
2130206	Promotion of forestry technology	√		
2130207	Forest resource management	√		
2130208	Forest resource monitoring	√		
2130209	Compensation for forest ecological benefit	√		
2130211	Animal and plant protection	√		
2130213	Forestry law enforcement and supervision	√		
2130216	Forestry quarantine and inspection	√		
2130217	Sand prevention and control	√		
2130218	Forestry quality and safety	√		
2130219	Forestry engineering and project management	√		
2130221	Forestry industrialization	√		
2130224	Forestry policy development and publicity	√		
2130233	Forest insurance subsidy	√		
2130234	Forestry disaster prevention and reduction	√		
2130299	Other forestry expenditures	√		
21303	Water conservancy		√	
2130301	Administrative operation		√	
2130303	Organ service		√	
2130304	Business management of water conservancy industry		√	
2130305	Construction of water conservancy projects		√	
2130306	Operation and maintenance of water conservancy projects		√	

Con.

Item No.	Item	High Correlation	Moderate Correlation	Low Correlation
2130309	Water law enforcement supervision		√	
2130310	Soil and water Conservation		√	
2130311	Water resources conservation management and protection		√	
2130313	Hydrological forecasting		√	
2130314	Flood prevention		√	
2130315	Drought prevention		√	
2130316	Farmland water conservancy		√	
2130317	Water technology promotion		√	
2130331	Expenditure on water resources arrangements		√	
2130399	Other water expenditures		√	
21304	South – to – North water diversion		√	
2130401	Administrative operation		√	
2130405	Policy research and information management		√	
2130407	Preliminary work		√	
2130408	South-to-North water diversion technology promotion		√	
2130499	Other South-to-North water diversion expenditures		√	
21305	Poverty alleviation			√
2130501	Administrative operation			√
2130550	Poverty alleviation institutions			√
2130599	Other poverty alleviation expenditures			√
21306	Comprehensive agricultural development			√
2130601	Institutional operation			√
2130602	Land management			√
2130603	Industrialized management			√
2130699	Other agricultural development expenditure			√
21399	Other agriculture, forestry and water resources expenditures			√
2139999	Other agriculture, forestry and water resources expenditures			√

Annex IV. Correlation between the classification of general public budget expenditure items and climate (detailed)

Con.

Item No.	Item	High Correlation	Moderate Correlation	Low Correlation
214	Transportation expenses			√
21401	Highway and waterway transportation			√
2140101	Administrative operation			√
2140102	General administrative affairs			√
2140104	New road construction			√
2140105	Road reconstruction			√
2140106	Road maintenance			√
2140108	Road administration			√
2140112	Road transportation management			√
2140113	Road passenger and freight station (field) construction			√
2140123	Channel maintenance			√
2140127	Ship inspection			√
2140131	Maritime management			√
2140136	Waterway transportation management expenditure			√
2140199	Other highway and waterway transportation expenditures			√
21402	Railway transportation		√	
2140206	Railway safety		√	
2140208	Industry regulation		√	
2140299	Other railway transportation expenditures		√	
21403	Civil air transport			√
2140301	Administrative operation			√
2140302	General administrative affairs			√
2140304	Airport construction			√
21499	Other transportation expenditures			√
2149999	Other transportation expenditures			√
215	Resource exploration information and other expenditures			√
21501	Resource exploration and development expenditures			√
2150199	Other resources exploration expenditures			√

Con.

Item No.	Item	High Correlation	Moderate Correlation	Low Correlation
21605	Tourism management and service expenditure			
2160501	Administrative operation			
2160502	General administrative affairs			
2160504	Tourism promotion			
2160505	Tourism industry business management			
2160599	Other tourism management and service expenditures			
21999	Other expenditures Land, marine and meteorology weather and other expenditures			
	Land resources affairs			√
2200101	Administrative operation			√
2200103	Organ service			√
2200107	Social services on land resources			√
2200108	Land resources industry business management			√
2200113	Investigation of geological and mineral resources			√
2200120	Expenditure on special income of mineral resources			√
2200150	Cause operation			√
2200199	Other land resources expenditure			√
22002	Marine management affairs			√
2200208	Marine law enforcement supervision			√
2200214	Marine use charges			√
2200217	Uninhabited island use charge			√
2200250	Cause operation			√
22003	Mapping affairs			√
2200301	Administrative operation			√
2200304	Basic mapping			√
2200305	Aerial photography			√
2200350	Cause operation			√
2200399	Other surveying and mapping affairs expenditures			√
22004	Earthquake affairs			√
2200404	Earthquake monitoring			√

Annex IV. Correlation between the classification of general public budget expenditure items and climate (detailed)

Con.

Item No.	Item	High Correlation	Moderate Correlation	Low Correlation
2200405	Earthquake forecasting			√
2200406	Earthquake prevention	√		
2200407	Earthquake emergency rescue	√		
2200450	Earthquake industrial institutions			√
2200499	Other earthquake expenditures			√
22005	Meteorological affairs	√		
2200504	Meteorological institutions	√		
2200509	Meteorological services	√		
2200510	Maintenance of meteorological equipment	√		
2200599	Other meteorological expenditures			√

Annex V. Detailed items of general budget expenditure classified based on correlation with climate change mitigation and adaptation

Table 1 Detailed items of general budget expenditure with high climate correlation classified based on correlation with climate change mitigation and adaptation

Item No.	Item	Mitigation			Adaptation		
		High Correlation	Moderate Correlation	Low Correlation	High Correlation	Moderate Correlation	Low Correlation
21102	Environmental monitoring and supervision	√			√		
2110203	Construction Project EIA Review and Supervision	√			√		
2110204	Nuclear and radiation safety	√			√		
21103	Pollution prevention and control	√					
2110301	Atmosphere	√					
2110304	Solid waste and chemicals	√					
2110305	Radioactive sources and radioactive waste supervision	√					
2110307	Expenditure on sewage charges	√					
2110399	Other pollution prevention and control expenditures	√					
21104	Natural ecological protection	√					
2110401	Ecological protection	√					
2110402	Rural environmental protection	√					
2110403	Nature reserve	√					
21110	Energy saving and utilization	√					

Annex V. Detailed items of general budget expenditure classified based on correlation with climate change mitigation and adaptation

Con.

Item No.	Item	Mitigation			Adaptation		
		High Correlation	Moderate Correlation	Low Correlation	High Correlation	Moderate Correlation	Low Correlation
2111001	Energy saving and utilization	√					
21111	Pollution reduction	√					
2111101	Environmental monitoring and information	√					
2111102	Environmental law enforcement supervision	√					
2130119	Disaster prevention and relief	√					
2130135	Protection, restoration and utilization of agricultural resources	√					
21302	Forestry	√					
2130201	Administrative operation	√					
2130203	Organ service	√					
2130204	Forestry institutions	√					
2130205	Forest cultivation	√					
2130206	Promotion of forestry technology	√					
2130207	Forest resource management	√					
2130208	Forest resource monitoring	√					
2130209	Compensation for forest ecological benefit	√					
2130211	Animal and plant protection	√					
2130213	Forestry law enforcement supervision	√					
2130216	Forestry quarantine and inspection	√					
2130217	Sand prevention and control	√					
2130218	Forestry quality and safety	√					
2130219	Forestry engineering and project management	√					
2130221	Forestry industrialization	√					
2130224	Forestry policy development and publicity	√					

Con.

Item No.	Item	Mitigation			Adaptation		
		High Correlation	Moderate Correlation	Low Correlation	High Correlation	Moderate Correlation	Low Correlation
2130233	Forest insurance subsidy	√					
2130234	Forestry disaster prevention and reduction	√					
2130299	Other forestry expenditures	√					
2200406	Earthquake prevention	√					
2200407	Earthquake emergency rescue	√					
22005	Meteorological affairs	√					
2200504	Meteorological institutions	√					
2200509	Meteorological services	√					
2200510	Maintenance of meteorological equipment	√					

Table 2   Detailed items of general budget expenditure with moderate climate correlation classified based on correlation with climate change mitigation and adaptation

Item No.	Item	Mitigation			Adaptation		
		High Correlation	Moderate Correlation	Low Correlation	High Correlation	Moderate Correlation	Low Correlation
20104	Development and reform affairs		√			√	
2010401	Administrative operation		√			√	
2010402	General administrative affairs		√			√	
2010404	Strategic planning and implementation		√			√	
2010408	Price management		√			√	
2010450	Cause operation		√			√	
2010499	Other development and reform expenditures		√			√	
20602	Basic research		√			√	
2060201	Institutional operation		√			√	
2060203	Natural science fund		√			√	
2060204	Key laboratories and related facilities		√			√	

Annex V. Detailed items of general budget expenditure classified based on correlation with climate change mitigation and adaptation

Con.

Item No.	Item	Mitigation			Adaptation		
		High Correlation	Moderate Correlation	Low Correlation	High Correlation	Moderate Correlation	Low Correlation
2060206	Special basic research		√			√	
2060299	Other basic research expenditures		√			√	
21101	Environmental protection management affairs		√			√	
2110101	Administrative operation		√			√	
2110102	General administrative affairs		√			√	
2110103	Organ service		√			√	
2110104	Environmental protection publicity		√			√	
2110199	Other environmental protection management expenditures		√			√	
2120108	Planning and protection of key national scenic spots		√			√	
21202	Planning and management of urban and rural communities					√	
2120201	Planning and management of urban and rural communities					√	
21203	Public facilities of urban and rural communities					√	
2120399	Other public facilities expenditures of urban and rural communities					√	
21205	Environmental health of urban and rural communities					√	
2120501	Environmental health of urban and rural communities					√	
21303	Water conservancy		√				
2130301	Administrative operation		√			√	
2130303	Organ service		√			√	
2130304	Business management of water conservancy industry		√				

Con.

Item No.	Item	Mitigation			Adaptation		
		High Correlation	Moderate Correlation	Low Correlation	High Correlation	Moderate Correlation	Low Correlation
2130305	Construction of water conservancy projects		√				
2130306	Operation and maintenance of water conservancy projects		√				
2130309	Water law enforcement supervision		√				
2130310	Soil and water conservation		√				
2130311	Soil and water conservation		√				
2130313	Hydrological forecasting		√				
2130314	Flood prevention		√				
2130315	Drought prevention		√				
2130316	Farmland water conservancy		√				
2130317	Water technology promotion		√				
2130331	Expenditure on water resources arrangements		√				
2130399	Other water expenditures		√				
21304	South-to-North water diversion		√				
2130401	Administrative operation		√			√	
2130405	Policy research and information management		√			√	
2130407	Preliminary work		√			√	
2130408	South-to-North water diversion technology promotion		√				
2130499	Other South-to-North water diversion expenditures		√				
21402	Railway transportation		√				
2140206	Railway safety		√				
2140208	Industry regulation		√			√	
2140299	Other railway transportation expenditures		√				

Annex V. Detailed items of general budget expenditure classified based on correlation with climate change mitigation and adaptation

Table 3 Detailed items of general budget expenditure with low climate correlation classified based on correlation with climate change mitigation and adaptation

Item No.	Item	Mitigation			Adaptation		
		High Correlation	Moderate Correlation	Low Correlation	High Correlation	Moderate Correlation	Low Correlation
20105	Statistical information affairs						√
2010501	Administrative operation						√
2010505	Special statistical business						√
2010506	Statistical management						√
2010507	Special census activities						√
2010508	Statistical sampling survey						√
2010550	Cause operation						√
20106	Financial affairs						√
2010601	Administrative operation						√
2010602	General administrative affairs						√
2010603	Organ service						√
2010604	Budget reform business						√
2010605	Financial treasury business						√
2010606	Financial supervision						√
2010650	Cause operation						√
2010699	Other fiscal expenditures						√
2040299	Other public security expenditures			√			√
20601	Science and technology management affairs			√			√
2060101	Administrative operation			√			√
2060102	General administrative affairs			√			√
2060103	Organ service			√			√
2060199	Other expenditures on science and technology management affairs			√			√
20603	Applied research			√			√
2060301	Institutional operation			√			√
2060302	Social welfare research			√			

Con.

Item No.	Item	Mitigation			Adaptation		
		High Correlation	Moderate Correlation	Low Correlation	High Correlation	Moderate Correlation	Low Correlation
2060399	Other applied research expenditures			√			
20604	Technical research and development			√			
2060401	Institutional operation			√			√
2060402	Application technology research and development			√			
2060403	Industrial technology research and development			√			
2060404	Transformation and diffusion of scientific and technological achievements			√			
2060499	Other technical research and development expenditure			√			
20605	Technical conditions and services			√			
2060501	Institutional operation			√			
2060503	Technical conditions special projects			√			
2060599	Other technical conditions and service expenditure			√			
20606	Social science			√			
2060601	Social science research institutions			√			
2060602	Social science research			√			
20607	Popularization of science and technology			√			
2060701	Institutional operation			√			
2060702	Science popularization activities			√			

Annex V. Detailed items of general budget expenditure classified based on correlation with climate change mitigation and adaptation

Con.

Item No.	Item	Mitigation			Adaptation		
		High Correlation	Moderate Correlation	Low Correlation	High Correlation	Moderate Correlation	Low Correlation
2060703	Youth science and technology activities			√			
2060704	Academic exchange activities			√			
2060705	Science and technology museum			√			
2060799	Other science and technology popularization expenditures			√			
20608	Science and technology exchange and cooperation			√			
2060801	International exchange and cooperation			√			
2060899	Other expenditures on science and technology exchange and cooperation			√			
20699	Other science and technology expenditures			√			
2069901	Science and technology awards			√			
2069999	Other science and technology expenditures			√			
212	Urban and rural communities expenditure						√
21201	Urban and rural communities management affairs						√
2120101	Administrative operation		√				√
2120105	Formulation and supervision of engineering construction standards		√				√
2120106	Engineering construction management		√				√
2120107	Municipal public sector market supervision			√			

Con.

Item No.	Item	Mitigation			Adaptation		
		High Correlation	Moderate Correlation	Low Correlation	High Correlation	Moderate Correlation	Low Correlation
2120109	Housing construction and real estate market supervision			√			
2120110	Professional qualification registration and review			√			
2120199	Other expenditures on urban and rural communities management			√			
21206	Construction market management and supervision			√			
2120601	Construction market management and supervision			√			
213	Agriculture, forestry and water resources expenditure		√				
21301	Agriculture		√				
2130101	Administrative operation		√				√
2130102	General administrative affairs		√				√
2130103	Organ service		√				√
2130104	Cause operation		√				√
2130106	Science and technology transformation and promotion services		√			√	
2130108	Pest control		√				
2130109	Quality safety of agricultural products		√			√	
2130110	Law enforcement supervision		√			√	
2130111	Statistical monitoring and information services		√			√	
2130112	Agricultural industry management		√			√	
2130122	Agricultural means of production and technical subsidies		√				

Annex V. Detailed items of general budget expenditure classified based on correlation with climate change mitigation and adaptation

Con.

Item No.	Item	Mitigation			Adaptation		
		High Correlation	Moderate Correlation	Low Correlation	High Correlation	Moderate Correlation	Low Correlation
2130123	Agricultural production insurance subsidy			√			
2130124	Institutionalized and industrialized agriculture management			√			
2130125	Processing and marketing of agricultural products			√			
2130126	Rural public welfare undertakings			√			
2130152	Subsidies for college graduates to work at the grassroots			√			
2130153	Expenditure on grassland vegetation restoration			√			
2130199	Other agricultural expenditures			√			
21305	Poverty alleviation			√			
2130501	Administrative operation			√			
2130550	Poverty alleviation institutions			√			
2130599	Other poverty alleviation expenditures			√			
21306	Comprehensive agricultural development			√			
2130601	Institutional operation			√			
2130602	Land management			√			
2130603	Industrialized management			√			
2130699	Other agricultural development expenditures			√			
21399	Other agriculture, forestry and water resources expenditures			√			
2139999	Other agriculture, forestry and water resources expenditures			√			

Con.

Item No.	Item	Mitigation			Adaptation		
		High Correlation	Moderate Correlation	Low Correlation	High Correlation	Moderate Correlation	Low Correlation
214	Transportation expenditures			√			
21401	Highway and waterway transportation			√			
2140101	Administrative operation			√			√
2140102	General administrative affairs			√			√
2140104	New road construction			√			√
2140105	Road reconstruction			√			√
2140106	Road maintenance			√			
2140108	Road transportation management			√			
2140112	Road transportation management			√			
2140113	Road passenger and freight station (field) construction			√			
2140123	Channel maintenance			√			
2140127	Ship inspection			√			
2140131	Maritime management			√			
2140136	Waterway transportation management expenditure			√			
2140199	Other highway and waterway transportation expenditure			√			
21403	Civil air transport			√			
2140301	Administrative operation			√			√
2140302	General administrative affairs			√			√
2140304	Airport construction			√			
21499	Other transportation expenditures			√			
2149999	Other transportation expenditures			√			
215	Resource exploration information and other expenditures			√			

Annex V. Detailed items of general budget expenditure classified based on correlation with climate change mitigation and adaptation

Con.

Item No.	Item	Mitigation			Adaptation		
		High Correlation	Moderate Correlation	Low Correlation	High Correlation	Moderate Correlation	Low Correlation
21501	Resource exploration and development expenditures			√			
2150199	Other resources exploration expenditures			√			
	Land resources affairs			√			√
2200101	Administrative operation			√			√
2200103	Organ service			√			√
2200107	Social services on land resources			√			
2200108	Land resources industry business management			√			
2200113	Investigation of geological and mineral resources			√			
2200120	Expenditure on special income of mineral resources			√			
2200150	Cause operation			√			√
2200199	Other land resources expenditures			√			
22002	Marine management affairs			√			
2200208	Marine law enforcement supervision			√			
2200214	Marine use charges			√			
2200217	Uninhabited island use charge			√			
2200250	Cause operation			√			
22003	Mapping affairs			√			
2200301	Administrative operation			√			√
2200304	Basic mapping			√			√
2200305	Aerial photography			√			√
2200350	Cause operation			√			√
2200399	Other surveying and mapping affairs			√			√

Con.

Item No.	Item	Mitigation			Adaptation		
		High Correlation	Moderate Correlation	Low Correlation	High Correlation	Moderate Correlation	Low Correlation
22004	Earthquake affairs			√			
2200404	Earthquake monitoring			√			
2200405	Earthquake forecasting			√			
2200450	Earthquake industrial institutions			√			
2200499	Other earthquake expenditures			√			
2200599	Other meteorological expenditures			√			

References

1. 中华人民共和国财政部：《2016 年政府收支分类科目》，中国财政经济出版社，2015 年。
2. 刘尚希："不确定性：财政改革面临的挑战"，《财政研究》，2015 年第 12 期。
3. 刘尚希："关于实体经济企业降成本的看法"，《财政研究》，2016 年第 11 期。
4. 刘尚希、石英华、罗宏毅："去产能不应是目标"，《财政研究简报》，2017 年第 12 期。
5. 苏明、王桂娟等：《中国气候公共财政统计分析研究》，2015 年 3 月，UNDP。
6. 苏明、石英华、王桂娟、陈新平："中国促进低碳经济发展的财政政策研究"，《财贸经济》，2011 年第 10 期。
7. 傅志华、石英华、封北麟、于长革："'十三五'时期推动京津冀协同发展的主要任务"，《经济研究参考》，2015 年第 62 期。
8. 任泽平、张庆昌："供给侧改革去产能的挑战、应对、风险与机遇"，《发展研究》，2016 年第 4 期。
9. 韩国高："供给侧改革下我国去产能的现状、挑战与对策分析"，《科技促进发展》，2016 年第 5 期。
10. 刘桂环、张彦敏、石英华："建设生态文明背景下完善生态保护补偿机制的建议"，《环境保护》，2015 年第 11 期。
11. 石英华："按照治理现代化的要求构建多元化的生态补偿资金机制"，《环境保护》，2016 年第 10 期。

12. 石英华:"积极稳妥推行中期财政规划管理",《公共财政研究》,2015年1月。

13. Mark Miller, 2012. "Climate Public expenditure and Institutional Reviews (CPEIRs) in Asia Pacific Region – What We learnt", UNDP, www. aideffectiveness. org/Climate ChangeFinance.

14. Adelante, Hanh Le, 2015. "A methodological Guidebook, Climate Public expenditure and Institutional Review", UNDP.

15. Governance of Climate Change Finance Team, 2015 (UNDP Bangkok Regional Hub), "Budgeting for climate change: How governments have used national budgets to articulate a response to climate change, lessons learned from over twenty climate public expenditure and institutional reviews", UNDP.

16. Hanh Le, Kevork Baboyan, 2015 "The case study of: Bangladesh, Indonesia, Nepal and the Philippines", UNDP draft working paper.

17. Kit Nicholson, Thomas Beloe, Glenn Hodes, 2016. "Charting New Territory: A Stock Take of Climate Change Financing Frameworks in Asia – Pacific", UNDP.